HANDBOOK ON PRESSURIZED IRRIGATION TECHNIQUES

HANDBOOK ON PRESSURIZED IRRIGATION TECHNIQUES

Second Edition

A. Phocaides
FAO Consultant

United Book Prints

Reprint : 2021

First reprint in India, 2017

Published by:
United Book Prints
4806/24, Ansari Road,
Daryaganj,
New Delhi - 110 002
Tel.: 011-41511055
E-mail: unitedbookprints@gmail.com

Print: 2017

ISBN: 978-93-83692-81-1

Printed in India

Contents

List of figures

List of tables

List of boxes

Foreword

Water is essential for all socio-economic development and for maintaining healthy ecosystems. As population increases and development calls for increased allocations of groundwater and surface water for the domestic, agriculture and industrial sectors, the pressure on water resources intensifies, leading to tensions, conflicts among users, and excessive pressure on the environment. The increasing stress on freshwater resources brought about by rising demand and growing pollution worldwide, is of serious concern.

Increasing water productivity holds the key to future water scarcity challenges. Today, agriculture accounts for 70 percent of all water use globally, up to 95 percent in several developing countries. Adding to the pressures on agricultural use is the increased awareness of the instrumental value of water in maintaining environmental services. Increasing the efficiency of water use and enhancing agricultural water productivity at all levels of the production chains are becoming priorities in a growing number of countries.

A comprehensive approach to agricultural water productivity requires actions at all levels, from crops to irrigation schemes, and up to national and international economic systems. In particular, shifting to modern on-farm irrigation practices can contribute to a substantial increase in both water use efficiency and water productivity.

The objective of this handbook is to provide a practical guide on the use of pressurized irrigation techniques to farmers, irrigation technicians, and extension workers in the field. In this second edition, the handbook has been considerably revised, including new chapters on low-cost drip irrigation and pipe distribution systems for smallholders.

Acknowledgements

The handbook on pressurized irrigation techniques is a joint effort of the Water development and management Unit of FAO (NRLW) and of the International Programme for technology and research in irrigation and drainage (IPTRID).

The first edition of the handbook was published in 2001 and was prepared by Andreas Phocaides, irrigation technology consultant, with the assistance of Reto Florin, former Chief of the FAO Water Service and David Casanova, irrigation expert.

For its second edition, the handbook was completely revised, with the addition of several new chapters. The author was assisted in its preparation by Ines Beernaerts and Jean-Marc Faurs (FAO), and Virginie Gillet (IPTRID).

List of acronyms

ABS	Acrylonitrile butadiene styrene (thermoplastic material)
AMIT	Affordable micro-irrigation technologies
ANSI	American National Standards Institute
ASAE	Society for Engineering in Agriculture, Food, and Biological Systems (former American Society of Agricultural Engineers)
ASTM	American Society for Testing Material
BHP	Break horspower
BOD	Biochemical oxygen demand
BS	British Standards
CAMS	Computer aided management systems
CEN	European Committee for standardization
CIF	Cost insurance and freight
COD	Chemical oxygen demand
CP	Center pivot
CYS	Cyprus Standards
DIN	Deutsches Institut fr Normung (German standards).
DN	Nominal diameter
ECe	Electrical conductivity
ECiw	Electrical conductivity of irrigation water
ECw	Electrical conductivity of water
EN	European Standard
ESP	Exchangeable sodium percentage
ET	Evapotranspiration
ETc	Crop water requirements
ETo	Reference evapotranspiration
FAO	Food and Agriculture Organization of the United Nations
FC	Field capacity
FDS	Family drip system
FOB	Free on board
HDPE	High density polyethylene
IES	Irrigation equipement supply database
IPTRID	Internatioanl Programme for Technology and Research in Irrigation and Drainage
ISO	International Standards Organization
kc	Crop coefficient
LDPE	Low density polyethylene
LEPA	Low energy precision application

LR	Leaching requirements
NTU	Turbidity
PC	Pressure compensated
PDS	Pipe distribution irrigation system
PE	Polyethylene
PIP	PVC irrigation pipe
PN	Nominal pressure
PP	Polypropylene
PR	Pressure rating
PVC	Polyvinyl chloride
PVC-U	Polyvinyl chloride unplasticised (equivalent to uPVC)
O&M	Operation and maintenance
RSC	Residual sodium carbonate
SDR	Standard dimension ratio
SS	Suspended solids
Sa	Available moisture
SAR	Sodium adsorption ratio
SC	Saturation capacity
TC	Technical Committee
TDR	Time domain reflectometry
TDS	Total dissolved solids
uPVC	Unplasticised polyvinyl chloride
USDA	United States Department of Agriculture
WHO	World Health Organization
WP	Wilting point

CHAPTER 1: Introduction

In the Proceedings of the Consultation on Irrigation in Africa (Lomé, Togo, 1997) irrigation was defined as "the application of water supplementary to that supplied directly by precipitation for the production of crops".

Although clearly defined, irrigation has not been clearly identified and separated from the wide-ranging area of water development activities, such as major and minor constructions for water harvesting, storing, conveyance and allocation; the drilling of tube-wells; and pumping. Most of the efforts and investments made in many countries for irrigation development result in water resources development and very few in on-farm water use improvement.

The application of improved irrigation methods and techniques on small farms is expanding rapidly as a result of the increasing demand for higher irrigation efficiency, improved utilization of water and intensification and diversification of production.

An irrigation system consists of canals and structures to convey, regulate and deliver the water to the users. Two basic types of irrigation systems exist: open canal systems (Figure 1.1) and pressured piped systems. This book concentrates in the latter one.

Experience gained from many countries in arid and semi-arid zones has shown that pressure piped irrigation techniques are replacing successfully the traditional open canal surface methods at farm level.

For any queries please contact: **FAO-water@fao.org**

FIGURE 1.1 - Wasteful surface irrigation method

CHAPTER 2: Pressure piped irrigation techniques

PRESSURE PIPED IRRIGATION SYSTEMS

A pressure piped irrigation system is a network installation consisting of pipes, fittings and other devices properly designed and installed to supply water under pressure from the source of the water to the irrigable area.

The basic differences between traditional surface irrigation and piped irrigation techniques are:

- The water flow regime: With traditional surface methods the size of the stream should be large, while in pressure piped irrigation systems very small flows, even 1 m^3/h, can be utilized.
- The route direction of the flow: With traditional surface methods the irrigation water is conveyed from the source and distributed to the field through open canals and ditches by gravity following the field contours. The piped system conveys and distributes the irrigation water in closed pipes by pressure following the most convenient (shortest) route, regardless of the slope and topography of the area.
- The area irrigated simultaneously: With traditional surface methods the water is applied in large volumes per unit of area, while piped irrigation systems distribute the water at small rates over a very large area.
- The external energy (pressure) required: Traditional surface gravity methods do not need external energy for operation, while piped irrigation systems require a certain pressure, 2–3 bars, which is provided from a pumping unit or from a supply tank situated at a high point.

NETWORK LAYOUT

The pipelines that convey and distribute the irrigation water to the individual plots are usually buried, and are so protected from farming operations and traffic hazards. Offtake hydrants, rising on the surface, are located at various spots according to the planned layout. With surface methods the irrigation water can be delivered directly to the open ditches feeding the furrows or the basins.

In micro-irrigation and other complete systems, e.g. sprinkler, the hydrants are coupled with smaller manifold feeder pipelines placed along the edges of the plots. These feed the lateral irrigating lines which are laid along the plants rows perpendicular to the manifolds. The laterals are equipped with water emitters at frequent spaces and distribute uniformly the irrigation water to the plants under certain pressure.

There are many kinds of irrigation systems. However, a thorough examination of the various system layouts, the equipment and the principles in operation shows that the same approach is always employed from the planning procedure to their application and that all of them have most of their features and components parts in common.

In all piped systems the main component parts (Figure 2.1) are:

- the control station (head control unit);
- the mains and submains (pipelines);
- the hydrants;
- the manifolds (feeder pipelines);
- the laterals (irrigating pipelines) with the emitters.

FIGURE 2.1 - Scheme of a network layout.

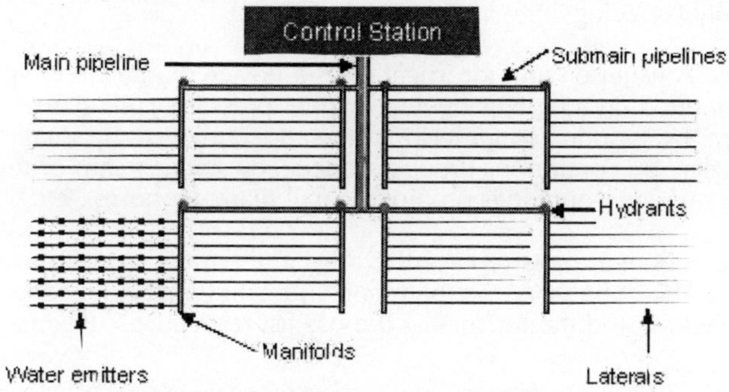

Head control. This consists of a supply line (rigid polyvinyle chloride or PVC, or threaded galvanized steel) installed horizontally at a minimum height of 60 cm above ground. It is equipped with an air release valve, a check valve, two ¾ inch hose outlets for connection with the fertilizer injector, a shut-off valve between the two outlets, a fertilizer injector and a filter. Where a gravel filter or a hydrocyclone sand separator is needed, it is installed at the beginning of the unit complex.

Main pipeline. It is the largest diameter pipeline of the network, capable of conveying the flow of the system under favourable hydraulic conditions of flow velocity and friction losses. The pipes used are generally buried permanent assembly rigid PVC, black high density polyethylene (HDPE), layflat hose, and quick coupling galvanized light steel pipes in sizes ranging from 63 to 160 mm (2–6 inches) depending on the area of the farm.

Submains. These are smaller diameter pipelines which extend from the main lines and to which the system flow is diverted for distribution to the various plots. The pipes are the same kind as the mains.

Offtake hydrants. These are fitted on the submains or the mains and equipped with a 2–3 inches shut-off valve. They deliver the whole or part of the flow to the manifolds (feeder lines).

Manifolds (feeder lines). These are pipelines of a smaller diameter than the submains and are connected to the hydrants and laid, usually on the surface, along the plot edges to feed the laterals. They can be of any kind of pipe available (usually HDPE) in sizes of 2–3 inches.

Laterals (irrigating lines). These are the smallest diameter pipelines of the system. They are fitted to the manifolds, perpendicular to them, at fixed positions, laid along the plants rows and equipped with water emitters at fixed frequent spacings.

Emitters. A water emitter for irrigation is a device of any kind, type and size which, fitted on a pipe, is operated under pressure to discharge water in any form: by shooting water jets into the air (sprinklers), by small spray or mist (sprayers), by continuous drops (drippers), by small stream or fountain (bubblers, gates and openings on pipes, small diameter hoses), etc.

These component parts replace the ones in the traditional surface systems, i.e. the main gate, the main and submain canals, the canal gates the field ditches, and the furrows or the basins, respectively (Figure 2.2).

FIGURE 2.2 - Improved surface irrigation method with pipes.

SYSTEM CLASSIFICATION

Piped irrigation systems are classified according to the pressure required for operation, the method of delivering water to plants, and the type of installation.

Pressure

The pressure of the system is the maximum water pressure required for normal system operation and encompasses: a) the friction losses in the piping network from the control station to the distal end of the system; b) the pressure required at the emitter; and c) the difference in elevation (plus or minus). Systems can be classed as:

- low pressure systems, where the pressure required is 2.0–3.5 bars;
- medium pressure, where the pressure required is 3.5–5.0 bars;
- high pressure, where the pressure required exceeds 5.0 bars.

Water delivery method

The water delivery method is the way the water is distributed to the plants. Systems can be classed as:

- *Sprinkler (overhead) irrigation:*The water is delivered in the form of raindrops precipitated over the entire area. There are many variations of this method in terms of the discharge and diameter coverage, the height of the water jet above ground (overhead, under the foliage), the type of sprinkler mechanism, etc.
- *Surface irrigation (furrow, basin, border, etc.):* The water is delivered to the field plots direct from the main or submain pipelines through the hydrants and it is spread all over the area, or it is side applied.
- *Micro-irrigation (localized irrigation)* by drippers, sprayers, bubblers, microjets, etc. The water is delivered to the plants without being spread over the entire area but by being applied in low rates to a limited soil surface area around the plants.

The water delivery method and the kind of the water emitter are the main characteristics of a piped irrigation system. In many cases they influence and specify the other characteristics (pressure and type of installation) and performances, such as the flow capacity of the system and the duration of application.

The flow capacity of a system is the water flow (in cubic metres per hour or litres per second) given, or designed to meet the irrigation requirements of the irrigable area at peak demand. It is inversely proportional to the duration of application. Where designed, it is usually the minimum permissible in order to economize on pipe size and other equipment. The duration of application is the time required for the completion of one irrigation cycle.

Type of installation

Systems can be classed as:

- Solid installations (fixed systems), where all the components are laid or installed at fixed permanent or seasonal positions.
- Semi-permanent installations, where the mains and submains are permanent while the laterals are portable, hand move or mechanically move.
- Portable installations, where all the component parts are portable.

PIPED IRRIGATION TECHNIQUES COMPARED WITH TRADITIONAL IRRIGATION METHODS

Irrigation efficiency. In open canal distribution networks, the water losses are estimated at up to 40 percent in unlined ditches and up to 25 percent in lined canals. These losses are due to seepage, phreatophytes and leakage in gates, spillways, etc. In piped systems, no such losses occur. During the application to the plants, the water losses range from 10 percent in localized micro-irrigation (Figure 2.3.) to 30 percent in overhead conventional sprinkler and surface methods (Figure 2.4). As a result, water losses can be minimized and an irrigation efficiency of 75–95 percent can be achieved. In open canals, the irrigation application efficiency ranges from 45 percent to a maximum of 60 percent.

Economic return per unit of water. Piped systems facilitate the manipulation of the irrigation water under more favourable conditions than do open canals. This can result in a yield increase of 10–45 percent and an improvement in quality.

Operation and maintenance (O&M). The man-hours needed in the piped systems range from one-tenth to one-quarter of those required for open canals. Any person can easily operate the piped systems, while the open canals can require skilled labour. In the open canals, expensive operations are carried out to prevent damage caused by roots; seepage through banks; the spread of weeds; siltation and sedimentation; clogging of outlets and gates; etc. In the piped systems, no maintenance or continuous repair of constructions is required. The basic component parts of the piped systems require minimal maintenance during the first seven years. The complete piped system requires a yearly maintenance costing about 5 percent of the initial investment.

Cost. The use of thermoplastic pipes and fittings, made of unplasticized polyvinyl chloride (rigid PVC), low density polyethylene (LDPE), high density polyethylene (HDPE), and polypropylene (PP), which are manufactured in almost every country in many sizes and classes, has reduced the cost of piped irrigation installations to a relatively low level at a time when open canal networks are becoming increasingly expensive.

The initial capital investment for the application of these techniques varies according to the method of irrigation and the type of the installation. The cost of the solid installations for localized methods is higher than that of the semi-portable hand-move sprinkler systems and the piped networks for surface methods. The costs for various piped irrigation systems installations in Europe are presented in Table 2.1 and the average percentage costs of the various parts of a piped system calculated on the basis for smallholdings (about 1.0 ha) are presented in Table 2.2. A detailed cost analysis of all the kinds and types of the piped systems has shown that the pipes (laterals included) account for about 50 percent of the total cost of the system.

The design complexity and the multiplicity of costly equipment is only apparent. The technology of piped irrigation systems is simple and flexible, and the investment yields a good return. Several mechanical difficulties are to be expected in the early stages. Subsequently, the farmers become familiar with the system's features and components and make the best use of it. The application of piped irrigation techniques produces a drastic change in irrigation management practices at farm level.

FIGURE 2.3 - Modern irrigation techniques.

FIGURE 2.4 - Sprinkler irrigation techniques.

TABLE 2.1 - Comparative costs of piped irrigation systems

	Piped surface method			Sprinkler conventional hand-move			Micro-irrigation solid installation		
Area (ha)	1	1-2	2-3	1	1-2	2-3	1	1-2	2-3
Installation cost (US$/ha)	1 700	1 600	1 400	2 800	2 700	2 100	3 950	3 300	3 000
Annual maintenance cost (US$/ha)	85	80	70	140	135	105	200	165	150

Note: Average 1997 prices in Europe.

TABLE 2.2 - Cost breakdown for piped irrigation systems

Component parts	Sophisticated installation	Simple installation
Control station	>23%	13%
Mains, submains and manifolds	10%	21%
Fittings and other accessories	22%	24%
Laterals (pipes and emitters)	45%	42%

CHAPTER 3: Irrigation equipment and jointing techniques

INTRODUCTION

Irrigation system installations consist of various pipes, fittings, valves and other equipment depending on the kind of system and the type of installation. Most installations have the same structure, and thus a relatively small range of equipment can meet the requirements of a whole region.

Irrigation equipment can be divided into:

- pipes;
- pipe connector fittings;
- flow control devices;
- filters;
- fertigation equipment;
- water emitters;
- automation equipment;
- operation equipment;
- water-lifting devices.

The main characteristics of the irrigation equipment are:

- material, e.g. galvanized steel, rigid PVC, etc.;
- size, i.e. the nominal diameter (DN) of the ISO metric range in millimetres (16–160 mm) and/or of the BSP threaded range in inches (3/4–4 inches);
- type of joint, e.g. threaded, quick coupling, solvent welded, etc.;
- working pressure PN (nominal pressure) or PR (pressure rating) in bars, e.g. 6.0 bars;
- national and/or international standards conformed to, e.g. DIN, ISO, BS, ASTM, EN.

The working pressure of a pipe or a fitting is the maximum internal water pressure to which the pipe or the fitting is subjected continuously in ordinary use, with certainty that failure of the pipe will not occur. It is specified as nominal pressure (PN) or pressure rating (PR).

PIPES

The pipes are the basic component of all irrigation networks. There are various kinds and types available in many pressure ratings and in different sizes (diameters). The pipes in use for farm-level irrigation systems are mainly in rigid PVC and polyethylene (PE). Quick coupling light steel pipes and layflat hoses are used on a smaller scale. Threaded galvanized steel pipes are of limited use. All these pipes are described below.

Steel threaded pipes. Galvanized steel pipes have been used widely in every country for all kinds of water works. In the past they were used as mains and submains in pressure piped irrigation solid installations. Due to their excellent properties, they have the ability to withstand stress, to resist high pressures and to maintain their strength for the duration of their service life, unlike plastic pipes which suffer a continuous creep strength with time and temperature fluctuations. They are not often used nowadays for irrigation because they are very expensive. However, they are useful in small pieces needed for risers in the hydrants, connector tubes in the head control units and similar applications. They are available in nominal diameters (DN), usually in inch-based series of 3/4, 1/2, 1, 1 1/4, 1 1/2, 2 inches, etc., which correspond more or less to the actual bore diameter, and in several high pressure rates (classes) in accordance with various standards and recommendations (ISO R-65, BS 1387, DIN 2440/41/42, or to American Standards, etc.). Supplied in random lengths of 6 m, they are for permanent assembling with screw-type (threaded) joints. Each pipe carries an internal threaded socket. Welded hot-dip galvanized steel pipes have an average life of 15–20 years on the surface 'in the atmosphere' and of 10–15 years in soil depending on soil physical properties. There is a large range of pipe connector fittings made of galvanized malleable iron for jointing these pipes (Figure 3.1).

FIGURE 3.1 - A threaded steel pipe fitting (male adapter).

Quick coupling light steel pipes. These pipes are made of light rolled strip steel which has been hot-galvanized inside and outside. Each pipe is equipped with a hand-lever quick coupling welded on one end while the other end is arranged accordingly for water and pressureproof tight closure. The standard pipe length is 6 m and the working pressure (PN) ranges from 12.0 to 20.0 bars. They are light in weight, easy to install and remove, and they are used as mains, submains, manifold feeder lines and laterals with sprinklers. They have a full range of pipe connector fittings of the same type of joints. They are available in many sizes and in diameters (DN) of 70, 76 and 89 mm, which are convenient for farm-level pressure irrigation techniques (Figure 3.2.).

FIGURE 3.2 - Quick coupling light steel galvanized pipes and fittings.

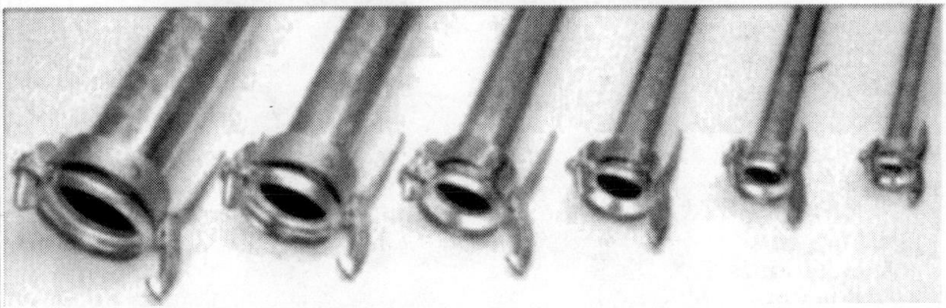

Quick coupling aluminium pipes. They are mostly used, always above ground, as moveable lateral lines in sprinkler irrigation portable installations. Made of aluminium alloy by extrusion or by fusion welding, they are light in weight (about half that of the light steel ones), relatively strong and durable. In accordance with ASAE S263.2, they are manufactured in nominal diameters quoted in inches, corresponding to the outside pipe diameter, of 2, 3, 4, 5 and 6 inches (51, 76, 102, 127 and 159 mm). The minimum working pressure is 7.0 bars. In accordance with ISO 11678, the same sizes in the metric series are 50, 75, 100, 125 mm and so on with working pressures of 4.0, 10.0 and 16.0 bars. They are supplied in standard lengths of 6, 9 and 12 m, complete with aluminium quick couplings. These are either detachable by means of clamps and rings, or permanently fixed on the tubes. With the use of U-shaped rubber gaskets, the couplings seal automatically under high water pressure during operation and drain in pressures below 1.0 bar. There are several types of quick couplings which allow the farmer to couple or uncouple the connections from any location along the pipe. The most widely used are the latch system (single or dual), with a ¾ or 1 inch threaded outlet for sprinkler risers, or hose extensions (Figure 3.3). Quick coupling provides a high degree of flexibility to aluminium pipelines laid on uneven ground. The expected life of these pipes is 15 years under good management. The light portable quick coupling pipes, steel or aluminium, can be used not only as sprinkler lateral lines, but also as water conveyance and distribution lines. In micro-irrigation systems they are often used as manifolds. These pipes maintain their value for a considerable length of time. Indeed, some cases have been reported of farmers selling many of these pipes at a profit even after extensive use.

FIGURE 3.3 - Quick coupling aluminium pipes.

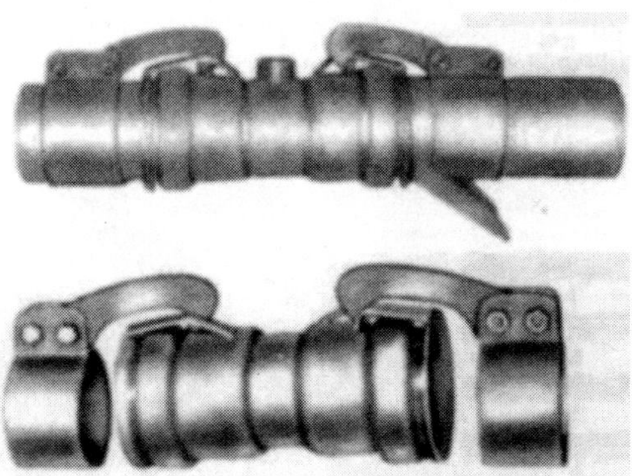

Rigid PVC pipes. Extruded from unplasticized polyvinyl chloride, also called uPVC for unplasticised polyvinyl chloride (or PVC-U), these pipes are ideal for irrigation, (cold) water conveyance and distribution lines as mains and submains. In many cases they can also serve as manifolds and laterals. Very light in weight, they are easy to transport and to handle on site. Their only limitations are that they must always be laid permanently underground, protected from high or very low ambient temperatures and solar radiation. The maximum flow velocity should not exceed 1.5 m/s. They are manufactured in standard lengths of 6 m, and in several series and classes denoting the working pressure, in accordance with various national and international standards applied in Europe, the United States and elsewhere (ISO 161-1/2: 1996, ISO 3606, BS 5556, DIN 8062, ASTM D 2241, ANSI/ASAE S376.1, ANSI/ASTM D 1785).

These standards, although equivalent to each other, vary in the pipe dimensioning, i.e. the pipe's actual diameter, the working pressure (PN), the safety factors, etc. In the United States, thermoplastic pipes are mainly classified in terms of standard dimension ratio (SDR) between the pipe's outside diameter and the pipe wall thickness, and schedules (for higher pressures). In Europe, the hydrostatic design stress (hoop strength) of PVC common material is 100 bars. In the United States, several compounds are used with different stress values, thus a great variety of pipes are produced, all in inch sizes. In accordance with the European standards and ISO 161, rigid PVC pipes are available in nominal diameters (DN), which is the approximate outside diameter, in 50, 63, 75, 90, 110, 125, 140, 160, 200 and 225 mm (Figure 3.4.). The working pressures are 4.0, 6.0, 10.0 and 16.0 bars at 24°C. At higher temperatures, the working pressures decrease accordingly. Usually, small diameter pipes up to 50 mm and inch-sized pipes have one end plain with a preformed socket at the other end for solvent cement welding. Larger diameter pipes have a tapered spigot at one end while the other end consists of a wall-thickened, preformed grooved socket with a rubber sealing ring for a push-fit integral mechanical joint.

There is a complete range of connector fittings for these pipes; some made of uPVC and others of cast iron. The compression-type polypropylene (PP) fittings are also suitable for uPVC pipes up to 110 mm. All the fittings and the valves of underground PVC pipelines should be thrust blocked to prevent them from moving whilst in operation due to the thrusting force of the water pressure. The estimated average life of buried uPVC pipes is 50 years.

Rigid PVC pipes are made for underground installation, where they are protected from temperature changes and hazards imposed by traffic, farming operations, etc. The trench should be as uniform as possible, firm, relatively smooth and free of large stones and other sharp edged material. Where ledge rock or hardpan is encountered, the trench bottom should be filled with embedment material, such as compacted grained soil or sand, to provide a bed depth of about 10 cm between pipe and rock. The minimum depth of cover should be 45 cm for pipes up to 50 mm, 60 cm for pipes up to 100 mm, and 75 cm for pipes over 100 mm DN. Where rigid PVC pipes are installed under roads, the depth of cover should not be less than 1 m; otherwise the pipes must be sleeved in a protective steel tube.

FIGURE 3.4 - Rigid PVC pipes.

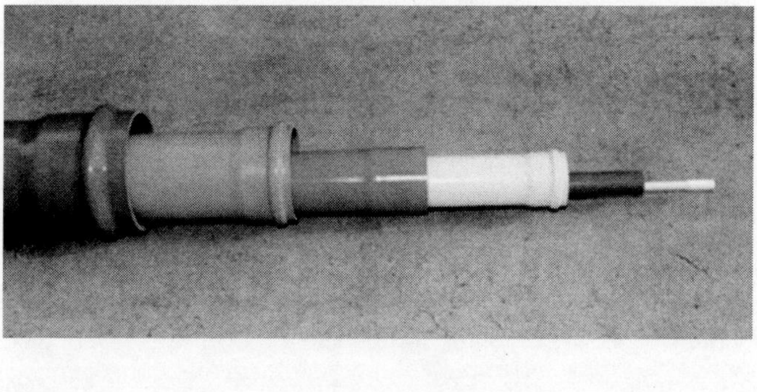

Polyethylene (PE) pipes. Flexible black PE pipes are extruded from polyethylene compounds containing certain stabilizers and 2.5 percent carbon black which protect the pipes against ageing and damage from sunlight and temperature fluctuations. LDPE (low-density resin) pipes are also known as soft polyethylene and PE 25, while HDPE pipes (highdensity resin) are more rigid and known as hard polyethylene or PE 50 (the numbers correspond to the pipe material's hydrostatic design stress). They are manufactured in accordance with various standards in inch-based and metric series (ISO 161-2, DIN 8072/8074, etc.) Both sorts have proved successful in pressure piped irrigation techniques and are the predominant kind of pipes in micro-irrigation systems. All laterals with micro-emitters are LDPE pipes (hoses) of 12–32 mm. HDPE pipes of larger diameters are used for main lines, submains and manifolds. They are also often used as water conveyance pipelines. LDPE pipes are less affected by high temperatures than HDPE pipes are. PE pipes are supplied with plain ends in coils of 50–400 m, depending on the diameter (Figure 3.5). Laid on the surface, they have a service life of 12–15 years. Conforming to European and international standards, they are available in the following sizes and working pressures:

DN (external diameter) millimetres:
12, 16, 20, 25, 32, 40, 50, 63, 75, 90 and 110;

PN (working pressure) bars:
2.0, 4.0, 6.0, 10.0 and 16.0.

Jointing PE pipes is simple. A full range of PP connector fittings is available in all diameters and types suitable for pressures from 2.0 to 10.0 bars.

FIGURE 3.5 - Polyethylene pipe in a coil.

The manufacturers of PVC and PE pipes recommend that the maximum flow velocity in the plastic pipes should not exceed 1.5 m/s. Based on this recommendation, Table 3.1 presents the flow rates in various plastic pipes with a flow velocity of 1.7 m/s, which should be taken as the maximum permissible under normal operating conditions.

TABLE 3.1 - Maximum recommended flow in plastic pipes without outlets

Rigid PVC 6 bars (DIN 8062)	DN mm Inside d. mm m^3/h	63 59.2 17	75 70.6 24	90 84.6 34	110 103.6 51	125 117.6 66	160 150.6 109
HDPE 6 bars (DIN 8074)	DN mm Inside d. mm m^3/h	50 44.2 9	63 55.8 15	75 66.4 21	90 79.8 30	110 97.4 45	
LDPE 4.0 bars DIN(8072)	DN mm Inside d. mm m^3/h	16 12.4 0.75	20 16.4 1.3	25 20.6 2.0	32 27.2 3.5		

V = 1.7 m/s

Selection of PVC and PE pipes's dimensions. Thermoplastic pipes (uPVC, PE etc.) have had widespread application in water conveyance, supply and irrigation in recent years. Many national and international special standards and specifications have been issued and many others are in the course of preparation for these pipes, in which different regulations and recommendations are included regarding the wall thickness of the various pipes as related with the pressure rating and the service time. The wall thickness s of the pipes is calculated by means of the equation given in ISO/R 161:

$$s = p\,\frac{d}{\sigma + p}$$

Where, $\boldsymbol{p}$ is the pressure rating, $\boldsymbol{d}$ the outside diameter of the pipe, and $\boldsymbol{\sigma}$ hydrostatic design stress i.e. the permissible stress of material in calculating the wall thickness $\mathbf{s}$. The value of $\mathbf{s}$ was based on a specific span of time creep rupture stress, making allowance for a safety factor too. For many years the manufacturers offered plastic pipes designed for standard values materials (PVC 100, HDPE 50 and LDPE 25). The hydrostatic design stress (creep strength) values given in kPa were based on a 50-year creep stress (service life) with safety factor 2.5 and maximum working (internal) pressure. These criteria in many cases were beyond the actual installation and operating conditions. Meantime new types of resins are available in the market with better mechanical strengths, higher than what has been required by current standards. The material designations have been updated, the safety factors modified and the wall thickness specifications have been amended to conform to ISO 4065. The pipe nominal pressure (PN) is no longer the basis for specifying the dimensions.

In accordance with DIN 8062 (1988) the un-plasticised uPVC pipes series are as in the previous standards. In the new standards the uPVC pipes are categorized in series (S), nominal pressure (PN) and standard dimension ratio (SDR). The sizes of the pipes (outside diameter) have not changed. There are many series and each one corresponds to a PN and an SDR (Table 3.2). The ISO 4065 gives a universal number of series, which correspond to specific series of other standards (see the following tables).

TABLE 3.2 - uPVC Pipe series

Nominal pressure in Bars (PN)	PN 4	PN 6	PN 8	PN 10	PN 12.5	PN 16
Pipes series to DIN 8062	2	3	–	4	–	5
Pipes series to ISO 4422 (S)**	25	16.7	12.5	10	8	6.25
Standard Dimension Ratio (SDR)	51	34.4	26	21	17	13.5
Pipe size d (out.dia.)			Wall thickness s mm			
75	1.8	2.2	2.9	3.6	4.5	5.6
90	1.8	2.7	3.5	4.3	5.4	6.7
110	2.2	3.2	3.4*	5.3	5.3*	8.2
125	2.5	3.7	3.9*	6	6*	9.3
160	3.2	4.7	4.9*	7.7	7.7*	11.9

Note: Up to the sizes of 90 mm the safety factor 2.5 is taken for designation of the pipes to DIN and ISO. For larger sizes > 90 to ISO the safety factor is 2.0. Then for the pipes PN 10 and PN 16 the wall thickness to ISO is less than the ones to DIN in sizes >90.
** ISO 4422 contains additional series, (S 20–SDR 41–PN 5), (S 16–SDR 33–PN 6.3), (S 4–SDR 9–PN 25).

The pipe dimensions given above are designed for a service life of 50 years at 20° Celsius under the precise PN. The three factors are interrelated and vary accordingly, e.g. at higher temperatures both the service life and the permissible working pressure change accordingly and the same applies when the working pressure differs from the designed one. Informative tables are included in the standard papers and/or in the manufacturers catalogues.

Regarding irrigation pipes made from polyethylene (PE) the worldwide known LDPE pipes to DIN 8072 are still in production, however new types of resins especially for high density (HDPE) are available in the market with better mechanical strengths, higher than what has been required by previous standards. Next to the familiar PE 25 and PE 50, there are the PE 32, PE 63, PE 80 and PE 100. The number corresponds to the σ value for the pipes designation. So the pipes for a specific PN differ in SDR and wall thickness according to the hydrostatic design stress of the row material made from. The selection of the appropriate pipe has become more accurate for the designed application, although more complicated (Table 3.3).

TABLE 3.3 - PE Pipes (50, 63, 80) wall thickness for PN 6 and PN 10

	PN 6			PN 10		
	Hydrostatic design stress σ (kPa)					
	PE 50	PE 63	PE 80	PE 50	PE 63	PE 80
	Pipe series (s) to ISO					
	S 8.3	S 10	–	S 5	S 6.3	S 8
	Standard Dimension Ratio (SDR)					
	17.6	21	–	11	13.6	17
Pipe d mm	Pipe wall thickness s mm					
50	2.9	–	–	4.6	–	–
63	3.6	–	–	5.8	4.7	–
75	4.3	–	–	6.8	5.6	4.5
90	5.1	4.3	–	8.2	6.7	5.4
110	6.3	5.3	–	10.0	8.1	6.6
125	7.1	6.0	–	11.4	9.2	7.4

Note: For the same PN with different material hydrostatic stress (σ) values there are different characteristics (SDR, S, wall thickness and weight).

Layflat hose. Layflat tubing has been used in irrigation for a number of years (Figure 3.6). It is an alternative to rigid PVC pipes for surface use as water conveyance lines, mains and manifolds, in drip and other low pressure micro-irrigation installations. It is made of soft PVC reinforced with interwoven polyester yarn. Layflat hoses are flexible, lightweight, and available in various sizes (millimetres or inches) from 1–6 inches and for working pressures (PN) of 4.0–5.5 bars. They are manufactured with plain ends and supplied in coils in standard lengths of 25, 50 and 100 m.

There are no special connector fittings for layflat hoses. The hoses are connected by inserting small pieces of PE piping into the ends of the hoses, or by metallic quick couplings attached to both pipe ends. Small diameter PE tubes are used to connect laterals to the layflat manifolds. In these cases, wire ties are needed to secure the connections. However, several micro-irrigation industries have designed and manufactured special connector fittings for jointing their drip lines with layflat hoses.

FIGURE 3.6 - A layflat hose.

PIPE CONNECTOR FITTINGS

Malleable iron threaded. These fittings are made for use with galvanized steel threaded pipes and they are available in a wide range as elbows, bends, reducers, tees, plugs, nipples and other (Figure 3.7). They are characterized by toughness and ductility and they provide a sound joint able to withstand pipeline expansion and contraction and other stresses. They are manufactured with screw type joints male and female (taper threads) in accordance with BS 21, DIN 2999, ISO R 7 and American standards in nominal sizes (inside diameters) as in the galvanized steel pipes. The sizes, usually quoted in inches, may be converted to millimeters i.e. ½ inch for DN 15 mm, ¾ inch for 20 mm, 1 inch for 25 mm, 1¼ inch for 32 mm, 2 inches for 50 mm, etc. Mostly they comply to BS 143&1256, DIN 2950, ISO R 49 for working pressures of minimum 14.0 bars.

Polypropylene (PP) pipe connector fittings. PP connector plastic pipe fittings (joints) are primary suitable for use with PE plastic pipes. There is a full range of all kinds, sizes and types of these pipe fittings worldwide.

FIGURE 3.7 - Threaded galvanized steel pipe fittings.

There are three main types with several modifications. These are:

- lock connector fittings, inserted into the pipe, with a locking ring which securely fastens the hose pipe to the fitting and can withstand high pressures;
- barbed type fittings, also inserted into the pipe, available only in small diameters up to 20 mm, and for pressures up to 2.0 bars only;
- compression type, which are available in all diameters and are for high pressures, 10.0 bars. The compression fittings are also suitable for larger size rigid PVC pipes. They are easily mounted and dismantled without cutting the pipe. They are more expensive than the other fittings but last longer and can be used in several installations.

All PP connector fittings are also available with one or both ends threaded (Figure 3.8).

FIGURE 3.8 - Polypropylene fittings compression type.

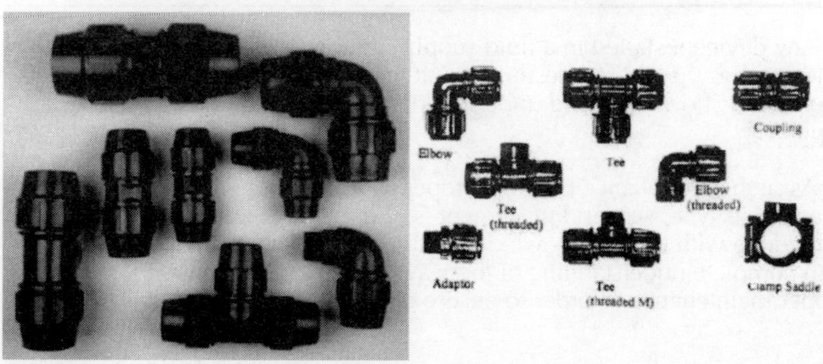

PVC fittings. Pipe fittings made of PVC are available in the inch system following the same dimensioning with the metal pipes and fittings and in the metric system (mm) conforming the ISO and DIN dimensioning. They are manufactured for solvent welding, threaded jointing and for push-fit integral mechanical jointing (Figures 3.9 and 3.10).

FIGURE 3.9 - PVC solvent welding pipe fitting.

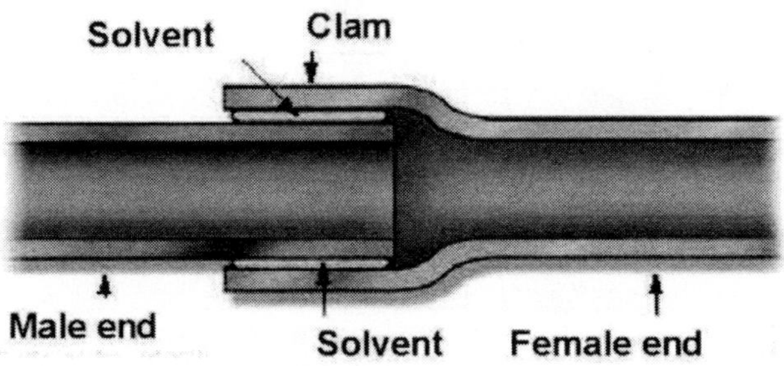

FIGURE 3.10 - PVC pipe with grooved ring socket push/fit joint

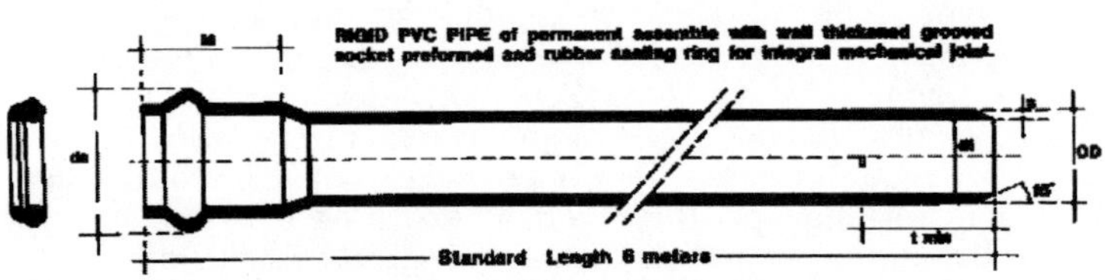

FLOW CONTROL DEVICES

Any device installed in a fluid supply system, in order to ensure that the fluid reaches the desired destination, at the proper time, in the required amount (the flow rate), and under the right pressure, is called a control appliance.

As such an appliance controls proper operation of a fluid system, selecting its type, size and placement is of uppermost importance and ought to be done with the full knowledge of the various features of the device and with complete understanding of the way it performs. Equally important is proper maintenance in order to ensure faultless and sound performance of the appliance. Made of metal base material or reinforced high engineering plastics, the common flow control devices used in irrigation systems can withstand high pressures (PN 10.0–16.0 bars) with screw-type end connections with internal or external threads for in-line installation.

Fluid control devices can be divided into three main classes (Table 3.4):

- Directional devices or valves. These serve to directly regulate the fluid flow. Installed in the pipeline, they enable starting or stopping the flow, and setting its rate, pressure and direction. Examples of such devices are the stop valves, the check valves and the regulating valves.
- Measuring devices or valves. In order to ensure the appropriate flow regime, just regulating the flow is not enough. It is also necessary to obtain accurate information about flow parameters, so that adjustments can be made, as required, to achieve the desired flow conditions. Water and flow meters and pressure gauges belong to this group.
- Auxiliary devices. These do not directly influence fluid flow, but ensure an undisturbed functioning of a system. To this group belong air valves and safety valves.

TABLE 3.4 - Scheme for flow control devices.

Directional devices or valves	Shut-off valves (stop valves)	• ball valves • butterfly valves • gate valves • disk* valves (globe, angle and oblique or Y valves) • radial valves
	Check valves (non-return valves)	• swing check valve • parallel check valve
	Regulating valves**	• disk* valves (globe, angle and oblique or Y valves) • radial valves
Measuring devices	Meters	• water meters • flow meters
	Gauges	• pressure gauges
Auxiliary devices	Air valves Safety valves	

Shut-off valves or stop valves. They are most widely used valves, manually operated (Figure 3.11). Usually installed between the ends of two pipes they serve to start or stop the flow of fluid in the pipeline. Stop valves are primarily designed for just two extreme situations: either to be completely open, to freely pass the full flow of fluid, or to be completely closed, to prevent any flow. The most common are the gate, ball, butterfly,

---

***:** *The sealing of a disk valves can be either a piston or a diaphragm.*
****:** *Regulating valves regulate pressure, flow or water level in either a direct acting or pilot operated way.*

FIGURE 3.11 - Various shut-off valves. From left to right a gate, a butterfly and ball valves.

radial and disk valves. In gate valves, the water moves in a straight line without resistance when fully open. Gate valves are not recommended for regulating or throttling flow, they must be either fully open or full closed. Ball valves are used on a large scale in small sizes of ¾ to 2 inches due to their many advantages. They feature quick on-off operation, quarter-turn and they can balance or throttle the flow. Of the disk valves, the most widely used model in irrigation networks is the oblique (Y-valve), ideal for throttling and regulating the flow. All types are made of brass in sizes of ½ to 4 inches, screw type with internal and/or external threads, at a PN of 16.0 bars. Oblique disk valves are also made of PP plastic material.

FIGURE 3.12 - Scheme and photograph of a check valve.

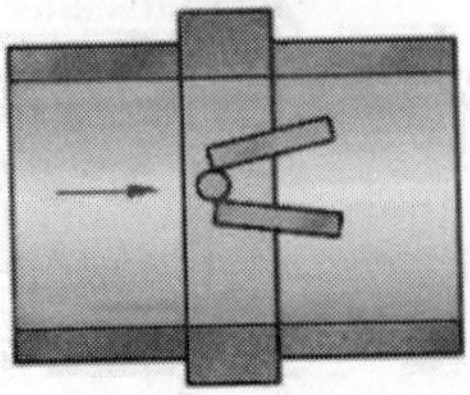

Check valves. Check valves, also called non-return valves, permit flow in one direction only and prevent reversal flow in piping by means of an automatic check mechanism (Figure 3.12). They come in two basic types: the swing check, which can be installed in horizontal or vertical piping; and the lift check, for use in horizontal lines only. Water flow keeps the check valves open, and gravity and reversal of flow close them automatically. They are placed in-line at the head control unit immediately after the pump. Swing checks are used with gate valves, lift checks with disk valves. Check valves are made of several metal materials and brass, and are screw type (female joints) quoted in inches from 3/4 to 4 inches, at a PN of 16.0 bars.

Regulating valves. Regulating valves are directional, semi-automatic devices, which allow controlling pressure and flow in a water supply system. These valves operate without any intervention from the operator, but the parameters of their performance must be preset by hand or by remote control, according to the requirements of the water supply system.

Regulating valves can be divided basically in three categories:

- For reducing the downstream pressure.
- For sustaining or relieving the upstream pressure.
- As flow regulators.

A pressure reducing valve will throttle flow and maintain downstream pressure at the required level, but only if the upstream pressure is higher than the preset level. Hence, it is controlled only by the downstream pressure (Figure 3.13).

A pressure sustaining valve will maintain upstream pressure either at its maximum permitted level, by relieving the superfluous flow, or at its minimum required level, by throttling the flow. Hence, it is controlled by the upstream pressure level.

Flow regulators are in-line valves that maintain a constant predetermined flow rate, regardless of pressure changes in the system.

All the types of regulating valves work on the principle of flow throttling, being based on the principle of inverse relationship between the cross-sectional area of a flexible orifice and the line pressure. In most cases the valves are of the disk (globe, angle and oblique or Y) or the radial type. Radial valves are often preferred to disk valves. Gate, ball, butterfly and other types are unsuitable for automatic regulation.

Regulating valves are either direct acting or pilot operated. Pressure regulating valves are often installed at the entrance of the manifolds to ensure a constant operating pressure for the laterals. They are made of brass, bronze or plastic in sizes of 1 to 3 inches with threaded connections.

FIGURE 3.13 - A pressure reducing valve maintains downstream pressure constant when pressure upstream increases.

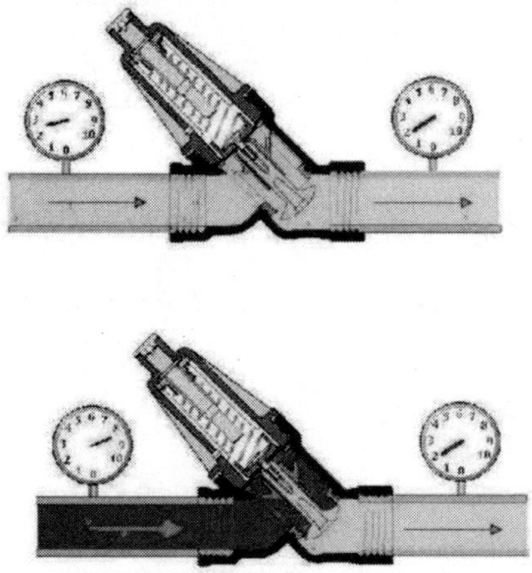

Meters. A distinction must be made between water meters and flow meters. Water meters measure and record the volumes of water passing through them, without considering the time element. Reading the output of a water meter gives information about the volume of water that passed through the appliance in a period, beginning with the last reading or zeroing of the meter. The most common type used for irrigation water is the Woltmann type with an impeller for axial flow. The velocity of flow activates the impeller and the turns are translated into total volume of water transmitted to the display dial through a series of reducing gears. They are manufactured in various designs, with the body made of cast iron, and constructed either as compact units or with an interchangeable inner mechanism. Sizes up to 2 in are available with threaded joints; larger sizes having flanges (Figure 3.14).

The flow meter measures the velocity of flow or, less often, the rate of flow or discharge. The most common type is the rotameter where a specially shaped float moves freely in a tube so that the flow velocity or rate is directly indicated by the float rim.

FIGURE 3.14 - Cross-section of a water meter.

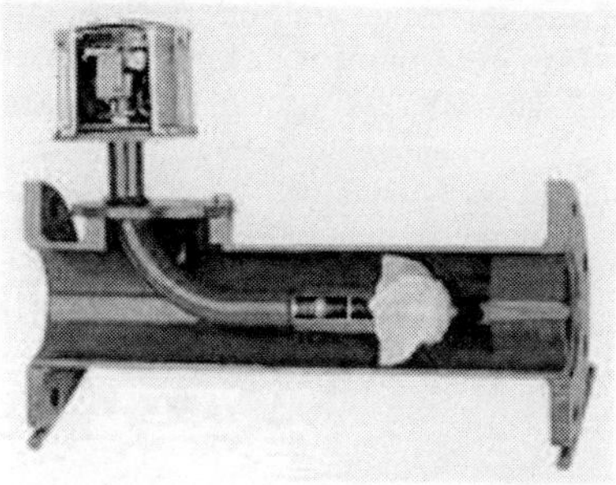

Pressure gauges. Measurement of pressure in key points of a network is of major importance for water system operator. The pressure gauge should be installed in easily accessible places, so that it is convenient to read and to maintain in proper working condition. The most common pressure gauge used in water supply and distribution service is the Bourdon gauge, in which the primary element is an elastic metal tube (Figure 3.15). As the pressure inside the tube increases the oval tube tends to become circular and this causes it to uncoil slightly.

FIGURE 3.15 - A Bourdon type pressure gauge.

Air valves. These valves are of great importance as they protect the pipe network from damage by trapped air in the system or from collapse due to a vacuum. If improperly chosen or located in a wrong place, it can also cause severe functional problems.

The presence of free air in water installations causes many difficulties in the piping system at start-up, during operation and when draining the system. Air valves are needed so that air can be either released from or admitted into the pipelines. Its operation and air flow rate cannot be influenced either by the system operator or by the performance of any other appliance (Figure 3.16). There are three main kinds of air valves:

1. Single automatic air release valves, for the continuous automatic release under pressure of the trapped air pockets accumulated at the summits of the mains. The single air valves are small in size with a 1 inch threaded connection, larger sizes not being required. They are installed on risers above ground at the high points of the conveyance and mains or every 200 m.

2. Large orifice air-vacuum valves (low pressure kinetic), for releasing or admitting air in bulk when filling or draining the system. They do not function under pressure. During normal operating conditions, a float held up by the system water pressure closes the large orifice. Sizes of 2 inches can meet the system requirements of 160 mm pipelines. They are installed at high points of the system after the pump or the system's main service hydrant on the head control unit, and at the beginning and the end of long branch pipelines.

3. Double (dual) air valves, which are a combination of the two above. They are the safest and most efficient air valves in mains and conveyance lines during filling, draining and operating the piped irrigation systems. The 2 inches size of the double air valve is appropriate for most on-farm piped irrigation installations up to 160 mm in diameter.

In addition to the above air valves, small vacuum breakers of inch are available for preventing vacuums in drip laterals laid on the soil surface, thus protecting them from clogging.

Air valves are manufactured for high working pressures of at least 10.0 bars PN. They are installed on-line with threaded internal or external joints.

FIGURE 3.16 - Air valves operate by introducing and releasing air to the water system.

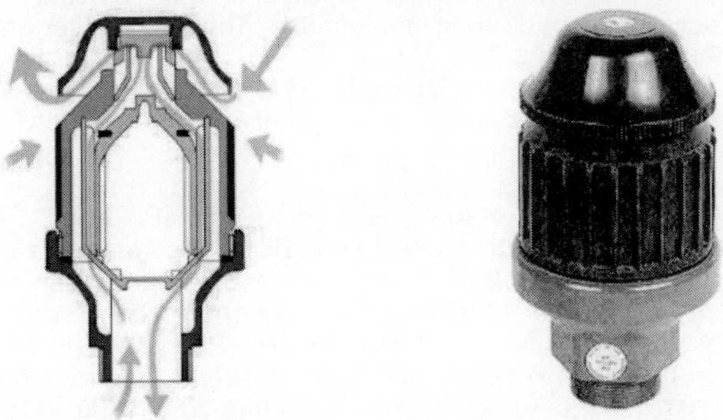

Safety valves (also called pressure relief valves). The practical use of safety valves began with steam boilers so that steam was released at critical pressures to avoid bursting of tanks and pipes. In water supply systems, the compressibility of water is very low and the problem of safety is therefore smaller. However, it is used mainly to ensure proper working of a system in cases of failure of other pressure control appliances.

Safety valves are on-line valves of smaller diameter than the pipelines, spring-loaded or otherwise, in which the outlet is inclined 90° to the inlet. When the pressure in the system exceeds the pre-set value, the valves open and release water into the air. Thus, they prevent the pipes from bursting due to sudden high pressures which might occur in the system. They are located immediately upstream of the main valve of the system. They are available in sizes of 1–3 inches with threads (Figure 3.17).

FIGURE 3.17 - Scheme and photograph of a safe valve.

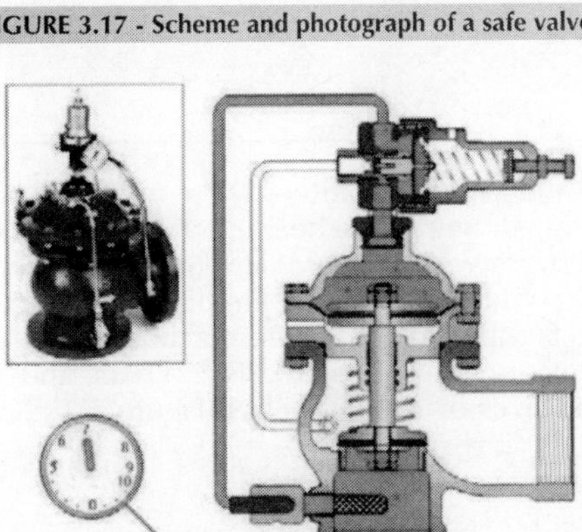

FILTERS

The filtration of the irrigation water is essential in order to avoid blockage damage to the micro-irrigation emitters. The type of filter used depends on the kind of impurities contained in the water and the degree of filtration required on the emitters. Their size should be the most economical with the lowest friction losses ranging from 0.3–0.5 bars. The following kinds of filters are available:

Gravel filters. These filters, also called media filters, are closed cylindrical tanks which contain a gravel grain of 1.5–3.5 mm or a basalt sand filter bed. Where the irrigation water source is an open reservoir, they are installed at the beginning of the head control of the system. Water entering the tank from top passes through the gravel bed, which traps the large particles of unbroken organic matter, mostly algae, and exits through the outlet at the bottom of the tank. They are equipped with the necessary inlet, outlet and drain valves, and a back-flushing arrangement. The filter body is epoxy coated metal, minimum 8.0 bars PN, and is 50–180 cm high and 40–100 cm in diameter. They are available in threaded connection sizes of 1 to 8 inches (Figure 3.18).

FIGURE 3.18 - Scheme and photograph of a gravel filter.

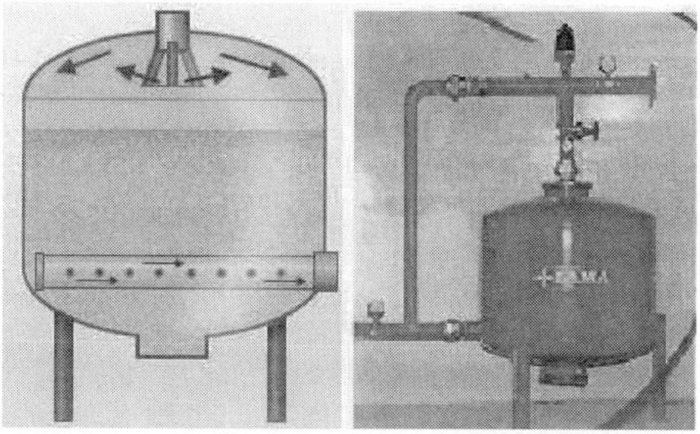

Hydrocyclone (sand separator) filters. These are closed conical metal tanks placed at the beginning of the head control unit where needed. They separate sand or silt from well or river water through the creation of a centrifugal force by a vortex flow inside the filter. This force drives the solids downward to a collecting chamber attached below and lets the clean water out. They are epoxy coated, PN 8.0 bars, and are available in threaded connection sizes of ¾ to 8 inches (Figure 3.19).

FIGURE 3.19 - Scheme and photograph of hydrocyclones.

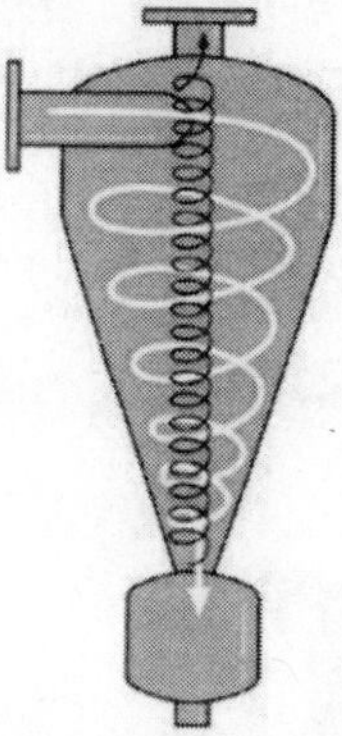

Screen type filters. These are used for final filtration as a safeguard for either moderate quality water or following a primary filtration with gravel or hydrocyclone filters. They are installed at the end of the head control before the main pipeline. They are made of epoxy coated metal or high engineering plastics in various cylindrical shapes (horizontal on-line, vertical angle, etc.), and are equipped with interchangeable perforated filtering elements, inlet, outlet and drain valves and pressure inspection gauges. They can withstand a working pressure (PN) of 8.0 bars. The degree of filtration ranges from 60 to 200 mesh (75 microns). They are available in sizes of ¾ to 4 inches. Smaller sizes are made of reinforced plastic (Figure 3.20).

FIGURE 3.20 - Operational scheme and photograph of a screen filter.

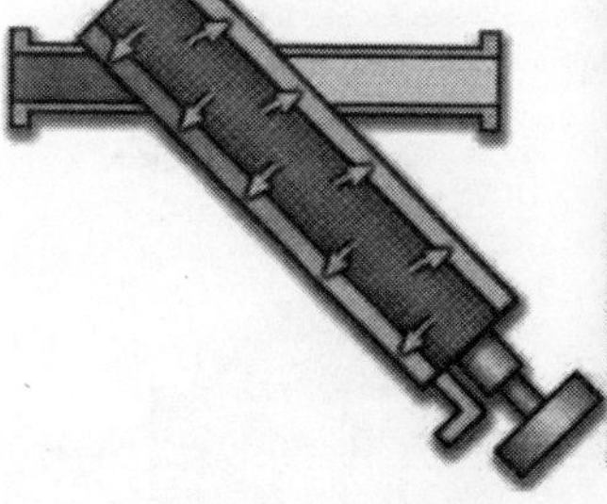

Disk type filters. They are cylindrical, made of reinforced plastic, horizontal in-line or vertical angle-shaped. The filtering elements consist of stacks of grooved plastic rings with multiple intersections providing a three dimensional filtration of high level. They are very effective in removing all kinds of impurities of inorganic and organic origin, algae included. The degree of filtration can range from 40 to 600 mesh (400–25 microns). They are available in all sizes (3/4 to 6 inches), PN 8.0 bars, with threaded joints. They are placed at the end of the control unit before the main pipeline (Figure 3.21).

FIGURE 3.21 - A grooved disk and a disk filter.

Automatic self-cleaning filters. Most of the different kinds and types of filters can be supplied with an automatic cleaning capability, as determined by pressure differential, duration of filtration, volume of water filtered, or by any combination of these. The cleaning mechanism, usually back flushing, for the removal of accumulated debris uses the system's water pressure. It is activated: a) whenever the pressure difference across the filter body increases to a predetermined value, e.g. 0.5 bar; and b) at fixed time intervals with an electronic timer back-up (Figure 3.22).

FIGURE 3.22 - An automatic self-cleaning filter uses the back flushing mechanism for the removal of accumulated debris.

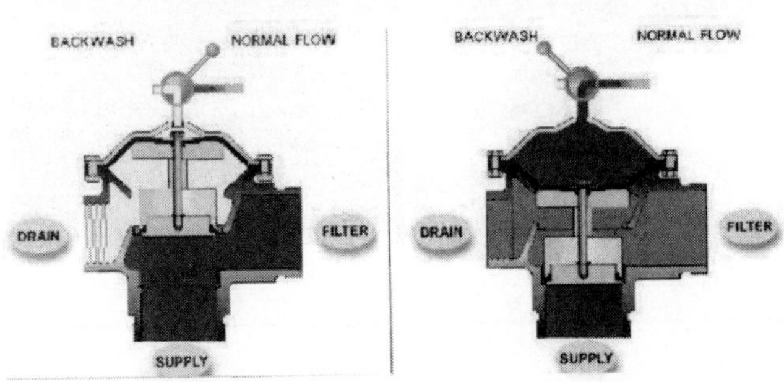

FERTIGATION EQUIPMENT

Fertilizers are applied with the irrigation water through the system using special devices called fertilizer injectors installed at the head control. There are three main types of fertilizer injectors: closed tank, Venturi type and piston pump. All of them are water driven by the operating pressure of the system.

Fertilizer (closed) tank. This is a cylindrical, epoxy coated, pressurized tank, resistant to the system's pressure, and connected as a bypass to the supply pipe of the head control. It is operated by differential pressure created by a partially closed valve, placed on the pipeline between the inlet and the outlet of the tank. Part of the flow is diverted to the tank entering at the bottom. It mixes with the fertilizer solution and the dilution is ejected into the system. The dilution ratio and the rate of injection are not constant. The concentration of fertilizer is high at the beginning and very low at the end of the operation. However, this apparatus is still in service on a very small scale in some countries because of its low cost and easy manufacture (Figure 3.23).

FIGURE 3.23 - Scheme and photograph of a fertilizer closed tank.

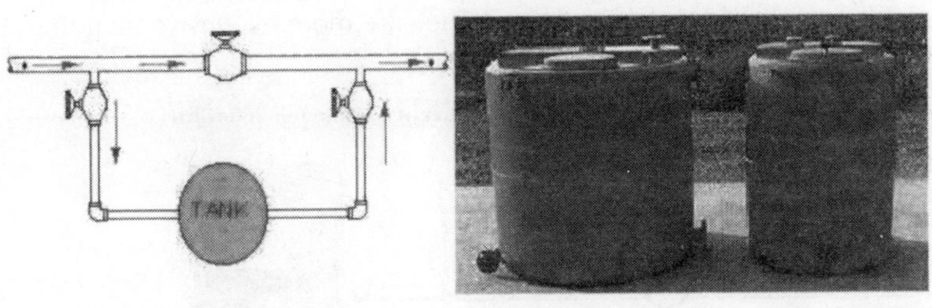

Venturi type. This is based on the principle of the Venturi tube. A pressure difference is needed between the inlet and the outlet of the injector. Therefore, it is installed on a bypass arrangement placed on an open container with the fertilizer solution. The rate of injection is very sensitive to pressure variations, and small pressure regulators are sometimes needed for a constant ejection. Friction losses are approximately 1.0 bar. The injectors are made of plastic in sizes from to 2 inches and with injection rates of 40–2 000 litres/h. They are relatively cheap compared to other injectors (Figure 3.24).

FIGURE 3.24 - Scheme and photograph of a Ventury type fertilizer.

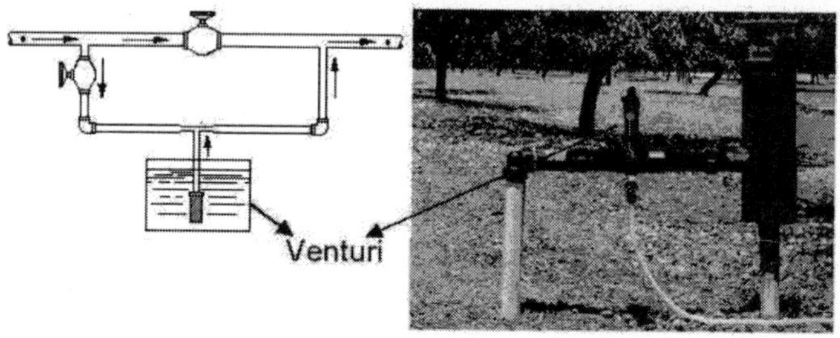

Piston pump. This type of injector is powered by the water pressure of the system and can be installed directly on the supply line and not on a bypass line. The system's flow activates the pistons and the injector is operated, ejecting the fertilizer solution from a container, while maintaining a constant rate of injection. The rate varies from 9 to 2 500 litres/h depending on the pressure of the system, and it can be adjusted by small regulators. Made of durable plastic material, these injectors are available in various models and sizes. They are more expensive than the Venturi type injectors (Figure 3.25).

FIGURE 3.25 - Scheme and photograph of a piston pump fertilizer.

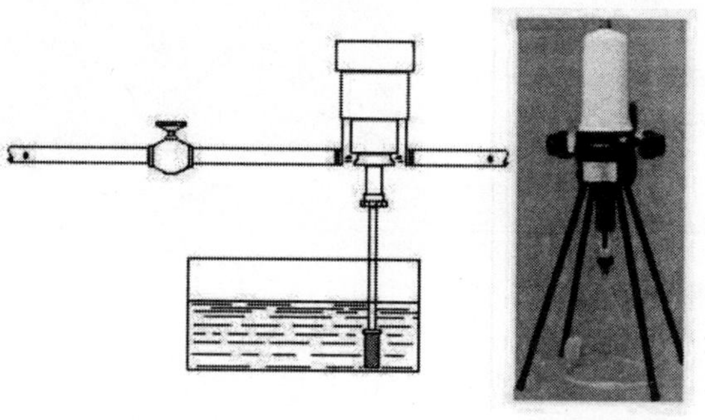

WATER EMITTERS

The water emitters specify the kind of system and in most cases the type of installation. Fitted on the laterals at frequent spaces, they deliver water to the plants in the form of a rain jet, spray, mist, small stream, fountain or continuous drops. All kinds and types of emitters in use now are of the small orifice-nozzle, vortex or long-path labyrinth types. Thus, the flow in the water emitters is turbulent. Some drip emitters of laminar flow used in the past are no longer available.

Sprinklers. Most of the agricultural sprinklers are the hammer-drive, slow rotating impact type, single or twin nozzle. The sprinklers shoot jets of water into the air and spread it to the field in the form of raindrops in a circular pattern. They are available in various nozzle sizes, flow discharges, operating pressures and wetted diameters or diameter coverage, full circle or part circle. They are classified as low, medium and high pressure/capacity, as shown in Table 3.5; according to the height of the water jet above the nozzle, they are divided into low angle (4°–11°), or high angle (20°–30°). They are made of brass or high engineering plastics with internal or external threaded connections of ¾–1½ inch. They are installed vertically on small diameter riser pipes, 60 cm above ground, fitted on the laterals. The sprinkler spacing in the field is rectangular or triangular at distances not exceeding 60 percent of their diameter coverage. Filtration requirements, where necessary, are about 20 mesh (Figure 3.26).

TABLE 3.5 - Sprinkler classification

Agriculture sprinklers (two nozzle)	Nozzle size mm	Operating pressure (bars)	Flow rate (m^3/h)	Diameter coverage (m)
Low pressure	3.0–4.5 x 2.5–3.5	1.5–2.5	0.3–1.5	12–21
Medium pressure	4.0–6.0 x 2.5–4.2	2.5–3.5	1.5–3.0	24–35
High pressure	12.0–25.0 x 5.0–8.0	4.0–9.0	5.0–45.0	60–80

FIGURE 3.26 - Rotating sprinklers.

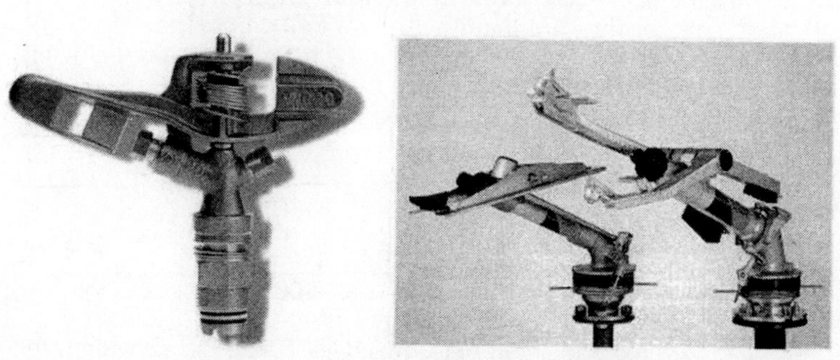

Microsprinklers. These water emitters are small plastic sprinklers of low capacity with flow rates less than 300 litres/h. Their main characteristics are their rapid rotation/whirling, less than a minute per rotation, the very small size of the water drops and the low angle of the water jet above nozzle. They have only one nozzle, of about 2.0 mm. They discharge 150–250 litres/h at 2.0 bars operating pressure. They are full circle and the wetted diameter is only 10–12 m. Mounted at a height of 60 cm on metallic or plastic rods inserted into the ground, they are connected to PE laterals (25 or 32 mm) through small flexible tubes 7 mm in diameter and 80 cm long. The spacing arrangement in the field is the same as for conventional sprinklers. The spacing does not exceed 6.0 m, i.e. 50 percent of the wetting diameter. The filtration requirements are about 60 mesh (300 microns) (Figure 3.27).

FIGURE 3.27 - A microsprinkler with a rotating spinner.

Spitters, micro-jets and sprayers. These are small plastic emitters with a low water discharge at a low angle in the form of fine drops in a sectorial or full circle pattern. They are mainly used for tree crops. They are of various mechanisms with a wide range of flow rates and water diameters. They have a small passage diameter, thus filtration of the water is essential. Their main performance characteristics are:

- operating pressure: 1.5–2.0 bars;
- flow rate: 35–250 litres/h (generally 150 litres/h);
- wetting diameter: 3–6 m;
- precipitation rate: 2–20 mm/h (generally 4–8 mm/h);
- filtration requirements: 60–80 mesh (250–200 microns).

Their heads are fixed to small plastic wedges 20–30 cm above ground and they are connected to the PE laterals with 7–9 mm flexible plastic tubes 60–120 cm long and a barbed plunger. They are placed one per tree, 30–50 cm apart (Figure 3.28).

FIGURE 3.28 - Photograph of a micro-jet irrigating a citrus tree and in detail the microjet's head with no-moving parts.

Bubblers. Low pressure bubblers are small-sized water emitters designed for localized flood irrigation of small areas. They deliver water in bubbles or in a low stream on the same spot. The flow rate is adjusted by twisting the top and ranges from 110 to 250 litres/h at operating pressures of 1.0–3.0 bars. The bubbler heads are installed, as are the minisprinklers, on small plastic wedges inserted into the ground and connected to a PE lateral with a 7 mm flexible plastic tube 80 cm long. They are placed in a tree basin; one or two per tree. The basin is always needed to contain or control the water because the bubbler discharge usually exceeds the soil infiltration rate (Figure 3.29).

FIGURE 3.29 - Scheme of a bubbler.

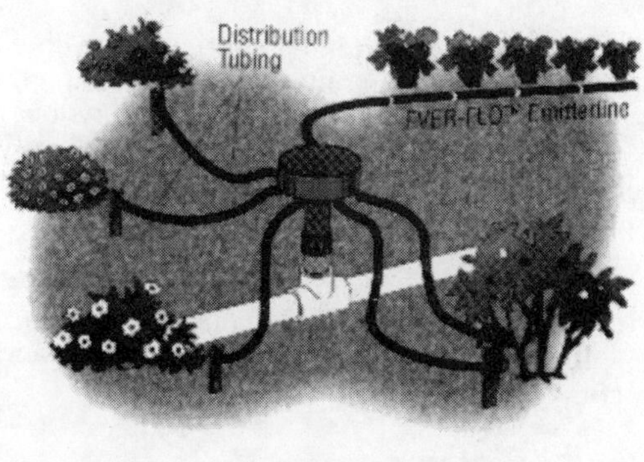

Drippers. The drippers are small-sized emitters made of high quality plastics. They are mounted on small soft PE pipes (hoses) at frequent spaces. Water enters the dripper emitters at approximately 1.0 bar and is delivered at zero pressure in the form of continuous droplets at low rates of 1.0–24 litres/h. Drippers are divided into two main groups according to the way they dissipate energy (pressure):

- orifice type, with flow areas of 0.2–0.35 mm^2;
- long-path type, with relatively larger flow areas of 1–4.5 mm^2.

Both types are manufactured with various mechanisms and principles of operation, such as a vortex diode, a diaphragm or a floating disc for the orifice drippers, and a labyrinthine path, of various shapes, for the long-path ones. All the drippers now available on the market are turbulent flow ones.

Drippers are also characterized by the type of connection to the lateral: on-line, i.e. inserted in the pipe wall by the aid of a punch; or in-line, where the pipe is cut to insert the dripper manually or with a machine (Figure 3.30).

On-line multi-exit drippers are also available with four to six 'spaghetti' type tube outlets.

FIGURE 3.30 - At left an on-line dripper, at right an in-line dripper.

Drip tapes. These are thin-walled integral drip lines with emission points spaced 10, 20, 30, 45 cm or any other distance apart, delivering lower quantities of water than the usual drippers at very low pressures, i.e. 0.4–1.0 litres/h at 0.6–1.0 bar. They are integrated drip lines where the drippers are built in the pipe walls at the desired spacing during the manufacturing process. They are ready-made dripper laterals with a very high uniformity of application. Drip tapes are made of LDPE or other soft PE materials in various diameters from 12 to 20 mm and in several wall thicknesses (0.10–1.25 mm). Thanks to a filtration system incorporated inside the tubing, they are less susceptible to mechanical and biological blockages than conventional drippers are (Figure 3.31).

FIGURE 3.31 - Scheme and photograph of a drip tape.

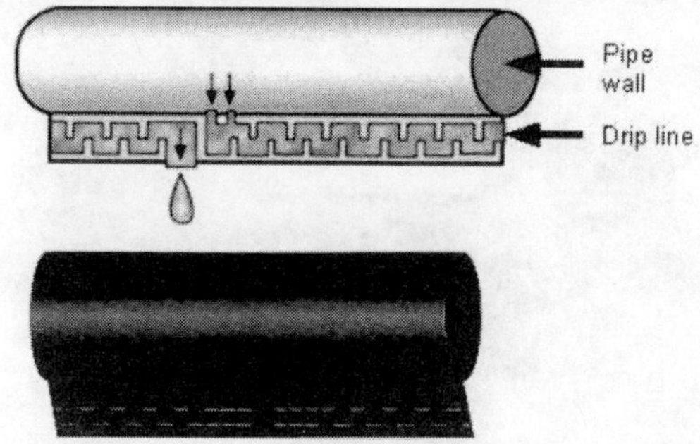

Pressure compensated (PC) emitters. Several sprinklers, drippers and other water micro-emitters are available with built-in flow regulators. These emitters deliver a constant flow of water at any pressure exceeding the fixed operating one. Uniform rates of discharge are achieved along the laterals regardless of the number of emitters, spacing, length of line or elevation, where excessive pressure is available. Therefore, pressure variations in the laterals due to friction losses can exceed 20 percent. Thus, less expensive smaller diameter pipes can be installed in certain cases. However, the self-regulated emitters, called pressure compensated, are normally operated under pressures exceeding the fixed operational pressures and cost more than the conventional ones (Figures 3.32 and 3.33).

FIGURE 3.32 - Differences in discharge between a normal emitter and an ideal and a real pressure compensated emitters.

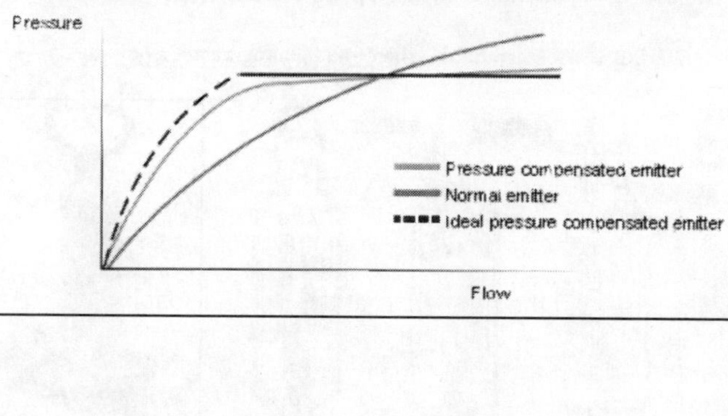

FIGURE 3.33 - Picture of a pressure compensated emitter showing the membrane that is used as flow-regulator.

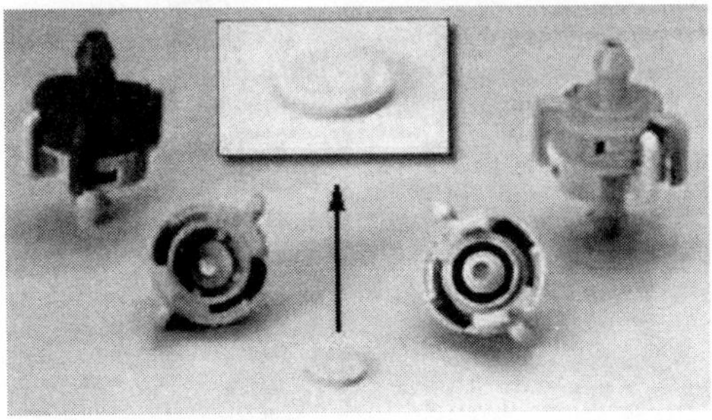

Pulsators. Pulsators are small plastic hydraulic devices used in micro-irrigation systems to reduce emitter and system flow rates to very low levels for higher efficiencies. The pulsators employ a built-in dripper with a discharge of 4–8 litres/h that feeds an integral silicone sleeve chamber. This in turn acts as a miniature pulsating pump generating hundreds of pulses per hour and so emitting the water. Thus, they can convert a low continuous flow into an instantaneous pressurized emission of water in short pulses. This process enables application rates of 0.3–0.8 mm/h with spitters, minisprinklers and sprayers, and 100–300 cm3/h with drippers. They are attached to the emitters, one for each minisprinkler or sprayer, and one for 20–70 drippers accordingly. The emitted water per pulse is roughly 0.5 cm^3. The silicone sleeve remains closed when the water pressure drops with the termination of the irrigation and prevents the water in the system from draining. Pressure compensated pulsators are also available for use on mountains and sloping terrain (Figure 3.34).

FIGURE 3.34 - Operational scheme and photograph of a pulsator.

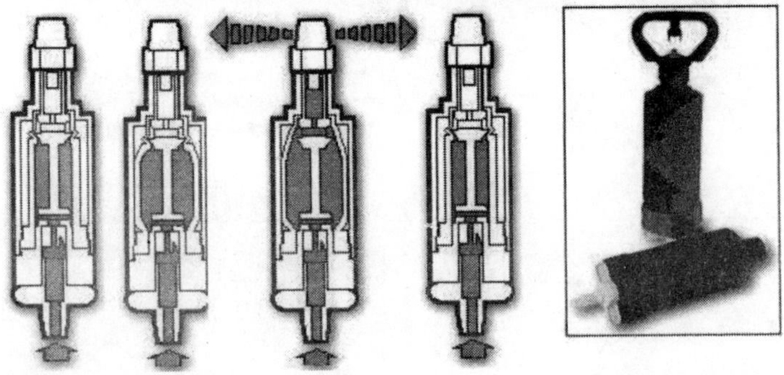

Porous pipes. These pipes are small-sized (about 16 mm) thin-walled porous flexible hoses made from PE fibres, PVC, ABS (Acrylonitrile Butadiene Styrene) or rubber. They permit water and nutrients under low pressure to pass from inside the tube, by transpiration, and irrigate the crops. The porous pipeline discharge is not accurate because the size of the pores varies and is not stable. They are used as lateral drip lines beneath the surface. Their application is limited although they do offer some advantages (Figure 3.35).

FIGURE 3.35 - Scheme of a porous pipe.

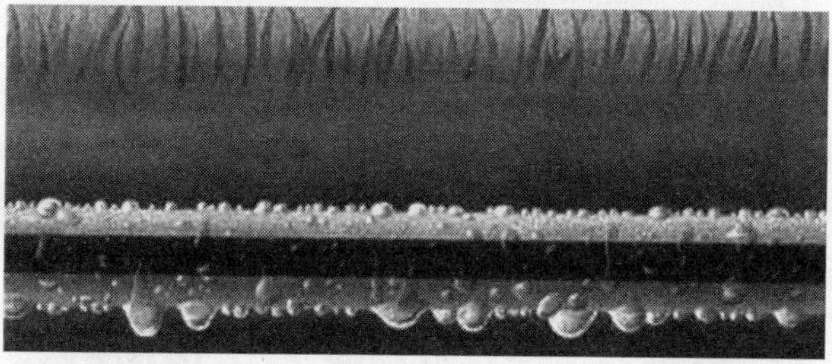

Garden hoses. Flexible garden hoses are made of various plastic materials, usually soft PVC, reinforced with textile or polyester yarn. They come in nominal diameters, the approximate inside diameter, of ½–1½ inch (15, 19, 25, 32 and 38 mm) with plain ends. They have a wide range of water applications (Figure 3.36).

FIGURE 3.36 - Photograph of a garden hose with outlet.

AUTOMATION EQUIPMENT

The main component parts for automation in an irrigation system are the remote control (electric) valves, the controller and the field wiring, where electricity is the transmitting power.

Electric (solenoid) valves. These are automatic valves which can be commanded from a distant point to turn the water flow on and off. The body construction is based on the globe valve design. They open and close by means of a flexible diaphragm or a piston utilizing hydraulic pressure controlled by an electrically actuated solenoid valve mounted on top (Figure 3.37). Made of reinforced glass or plastic, the electric valves, normally closed, are in inches with screw-type connections, a working pressure of 10.0 or 14.0 bars, and with a handle for manual operation and flow control.

FIGURE 3.37 - Operational scheme and photograph of an electric control valve.

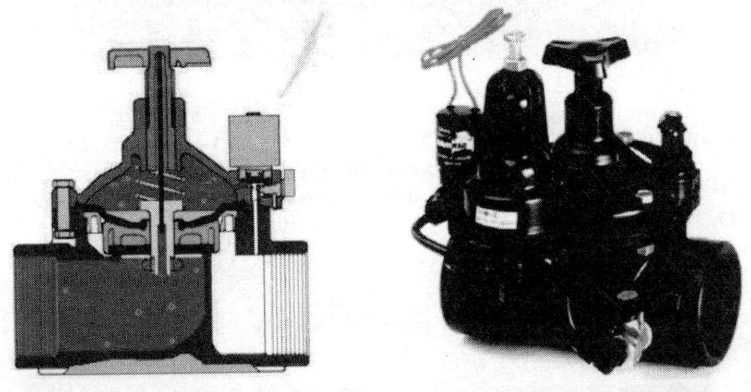

Controllers. These are automatic timing devices which supply the actuating power to operate the remote control (electric) valves, i.e. to open and close on a pre-set programme. They contain a transformer which reduces the standard voltage to 24–30 V. The power output from the electric controllers is transmitted to the electric valves through underground wiring. Their main features are the stations and the programmes. Each station usually operates one valve. The operation of the stations is sequential. There are many types of controllers available in many stations, up to 30, with dual or triple programmes for different scheduling and variable cycles of more than 14 days, and 0–12 hour station timing. Battery powered controllers are also available for independent stations (Figure 3.38).

FIGURE 3.38 - Operational scheme and photograph of an irrigation controller device.

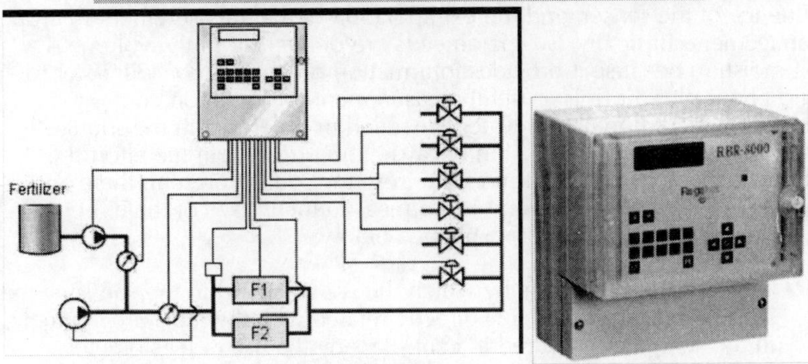

Automatic volumetric metering valves. These valves consist of a water meter, a pilot assembly and a shut-off mechanism. Once the pre-set volume of water has been delivered, they shut off automatically (Figure 3.39). Small sizes are operated mechanically and larger ones hydraulically by means of a diaphragm or a piston. Small sizes are available with screw-type joints while larger sizes, made of cast iron, come in flanged connections. They have a relatively limited application, mainly due to their high cost.

FIGURE 3.39 - Operational scheme and photograph of an automatic volumetric metering valve.

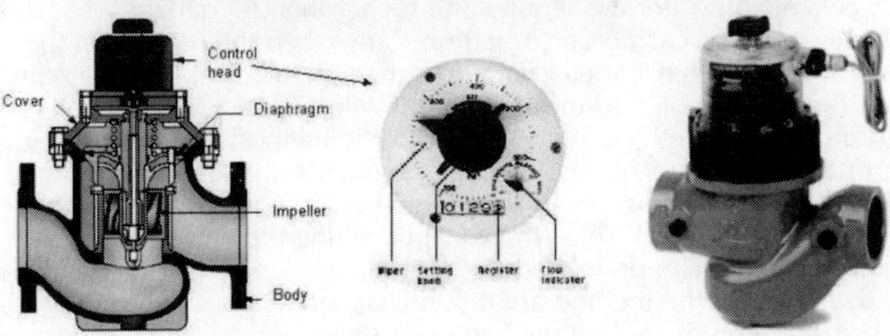

OPERATION EQUIPMENT

For the proper management of the irrigation systems, frequent simple water and soil checks and other measurements must be carried out on site. For this purpose, there are several instruments that give direct readouts of the results.

Soil moisture sensors. Soil moisture measurement is difficult mainly because of the variability in soil types, calibration of the sensor, area of influence of the sensor, and the extrapolation of that measurement to crop management. Basically, two parameters are of interest: (i) the volumetric soil moisture because it provides information on the ratio of soil water to solid phase plus air (it is a useful measurement for irrigation control informing about how much water is needed in order to "fill the sponge") and (ii) the soil moisture tension because it informs about the effort the plant has to make to extract moisture from the soil. At present, there are various technologies for soil moisture measurement. Two of them, because of their relevance, are briefly explained below:

- Tensiometers: The force by which the water is held in the soil is called soil water tension and it is directly related with the moisture content in the soil. It is measured in centi-Bars by the use of tensiometers. Nearly all types of tensiometers consist of three parts, a closed plastic tube, a ceramic porous cap at the bottom and a vacuum measuring gauge at the top. They are available for various depths in several lengths from 30 to 150 cm. They are filled with de-aired water and inserted permanently into cored holes in the soil near the plants, always in pairs, at two different depths, one at 30 percent the effective root-depth and one at 60 percent. Good contact of the surrounding soil must be arranged. The water availability is easily controlled and high osmotic pressure values in the root environment can be secured. At the state of field capacity the soil moisture tension is normally 10 cBars in sandy soils, 20 cBars in medium and 30 cBars in clays. Readings below 10 centibars indicate saturated soil, 20–40 excellent soil water availability and higher than 55 danger for severe water stress. In cash crops irrigation starts when the reading of the shallow depth tensiometer is in the range of 18–25 cBars depending on the type of soil and the stage of growth (Figure 3.40).
- Time Domain Reflectometry (or TDR). The method is based on the principle that velocity of an electromagnetic wave depends on the conducting medium. The larger the soil water content is, the slower the wave will travel. Thus, the wave travelling time along a probe of known length can be related to the soil water content. The main advantages of the method are that it is accurate, continuous, and it does not require calibration. The main disadvantage is that it has complex electronics and expensive equipment.

Conductivity meters. These portable instruments are battery powered and enable rapid and accurate determination of the concentration of soluble salts in the soil solution and the irrigation water. They are temperature compensated but they need frequent calibration (Figure 3.41).

Soil solution extractors. These instruments consist of a plastic tube with a porous ceramic cap at the bottom, like the tensiometers, and a syringe. They are inserted in pairs into the soil at the root plant area (in microirrigation

FIGURE 3.40 - Scheme and photographs of tensiometers and a TDR.

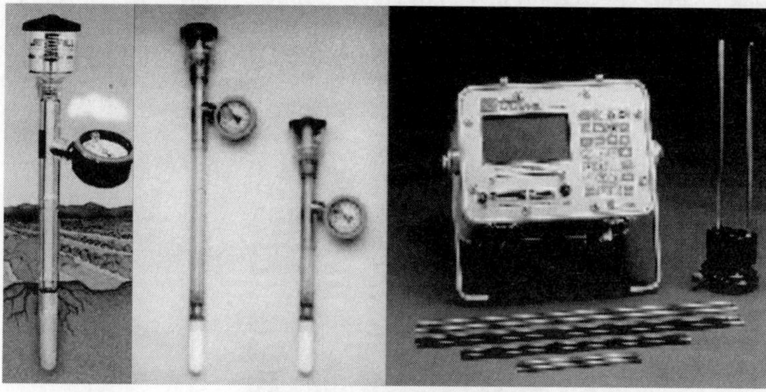

FIGURE 3.41 - Direct measurement of soil solution with a conductivity meter and extractors in the field.

one near the emitter and one between the emitter's lines). A vacuum is created within the empty tube and forces the moisture to move from the soil into the extractor through the ceramic cap. The solution collected is then withdrawn by the syringe for evaluation. These instruments enable the continuous follow up of the change in the soil total salinity, the chlorides and the nitrates content and the pH, as a result of intensive irrigation and fertigation. They are available in various lengths from 15 to 150 cm.

Class A evaporation pan. This is an open circular pan which is widely used to measure evaporation. It is made of 22-gauge galvanized steel plate, all surfaces painted aluminium, or of 0.8 mm Monel metal. The pan has a standard size: 121 cm in diameter and 25.5 cm deep. It is placed level on a wooden beamed base support 15 cm above ground (Figure 3.42). It is filled with water to 5 cm below the rim. It has a simple or advanced reading mechanism to indicate the decrease in water level due to evaporation. Measurements are recorded every morning at the same time. The water is topped up when its level drops to about 7.5 cm below the rim.

FIGURE 3.42 - A class A evaporation pan in the field.

WATER-LIFTING DEVICES

Pumps and lifting/propelling devices are often classified on the basis of the mechanical principle used to lift the water: direct lift, displacement, creating a velocity head, using the buoyancy of a gas or gravity. Most categories sub-divide into further classifications "reciprocating/cyclic" and "rotary". The first of these relates to devices that are cycled through a water-lifting operation (for example, a bucket on a rope is lowered into the water, dipped to fill it up, lifted, emptied and then the cycle is repeated); in such cases the water output is usually intermittent, or at best pulsating rather than continuous. Rotary devices were generally developed to allow a greater throughput of water, and they also are easier to couple to engines or other mechanical drive.

Virtually all water lifting devices can best be characterized for practical purposes by measuring their output at different heads and speeds. Normally the performance of a pump is presented on a graph of head versus flow (an H-Q graph, as in Figure 3.43) and in most cases curves can be defined for the relationship between **H** and **Q** at different speeds of operation. Invariably there is a certain head, flow and speed of operation that represents the optimum efficiency of the device, i.e. where the output is maximized in relation to the power input. Some devices and pumps are more sensitive to variations in these factors than others; i.e. some only function well close to a certain design condition of speed, flow and head, while others can tolerate a wide range of operating conditions with little loss of efficiency. For example, the centrifugal pump characteristic given in Figure 3.42 shows an optimum efficiency exceeding 80 percent is only possible for speeds of about 2000 rpm.

FIGURE 3.43 - Typical curves showing relationship between head, flow, speed, and efficiency (example for a centrifugal pump).

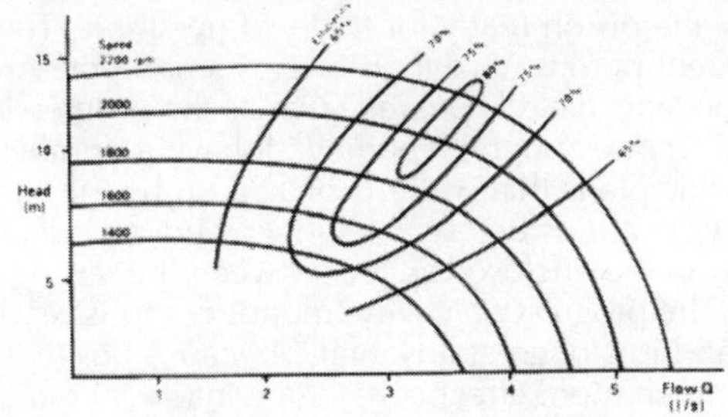

Direct-lift devices. These are all variations on the theme of the bucket and are the earliest artificial method for lifting and carrying water. It generally improves both efficiency and hence productivity if the water lifting element can move on a steady circular path. An obvious improvement to the simple rope and bucket is to fit numerous small buckets around the periphery of an endless belt to form a continuos bucket elevator. The original version of this, which is ancient in origin but still widely used, was known as a "Persian wheel". The water powered Noria, a water wheel with pots, is similar in principle (Figure 3.44).

FIGURE 3.44 - Scheme of a Persian wheel and photograph of a Noria.

Displacement pumps. The most common and well-known form of reciprocating/cyclic are the piston-type pumps and of rotary/continuous are the Archimedean screw-types. In the piston pump, water is sucked into the cylinder through a check valve on the up-stroke, and the piston valve is held closed by the weight of water above it; simultaneously, the water above the piston is propelled out of the pump. On the down-stroke, the lower check valve is held closed by both its weight and water pressure, while the similar valve in the piston is forced open as the traped water is displaced through the piston ready for the next up-stroke. The rotary positive displacement pumps have their origins among the Archimedean screw. Modern concepts have appeared such as the progressive cavity pump (also called "mono" pump), yet they all have a number of similarities. The principle is that water is picked up by the submerged end of the helix each time it dips below the surface and, as it rotates, a pool of water gets trapped in the enclosed space between the casing and the lower part of each turn. The progressive cavity (mono) pump is ready to fit in down boreholes and it is of great advantage because positive displacement pumps can cope much more effectively than centrifugal pumps with variations in pumping head. Therefore, any situation where the water level may change significantly with the seasons makes the progressive cavity pump an attractive option (Figure 3.45).

FIGURE 3.45 - Left: Scheme of a hand pump with single acting bucket piston (piston valve shown open as on the down-stroke, and inlet valve is closed). Center: A peadle piston-type pump. Right: Scheme of a progressive cavity or "mono" pump.

Velocity pumps. Their mechanism is based on the principle that when water is propelled to a high speed, the momentum can be used either to create a flow or to create a pressure. The reciprocating/cyclic ones are rarely used while the rotary/continuous ones are highly widespread. The latter are called rotodynamic pumps and their mechanism is based on propelling water using a spinning impeller or rotor.

Since any single rotodynamic pump is quite limited in its operating conditions, manufacturers produce a range of pumps, usually incorporating many components, to cover a wider range of heads and lows. Where high flows at low heads are required (which is common in irrigation pumps), the most efficient impeller is an axial-flow one (this is similar to a propeller in a pipe). Conversely, for high heads and low flows a centrifugal (radial-flow) impeller is needed (Figure 3.46).

Where a higher head is needed than can be achieved with a single pump, two pumps can be connected in series. Similarly, if a greater discharge is needed, two centrifugal pumps may be connected in parallel (Figure 3.47).

FIGURE 3.46 - Rotodynamic pumps: (a) radial or centrifugal pump and (b) axial pump.

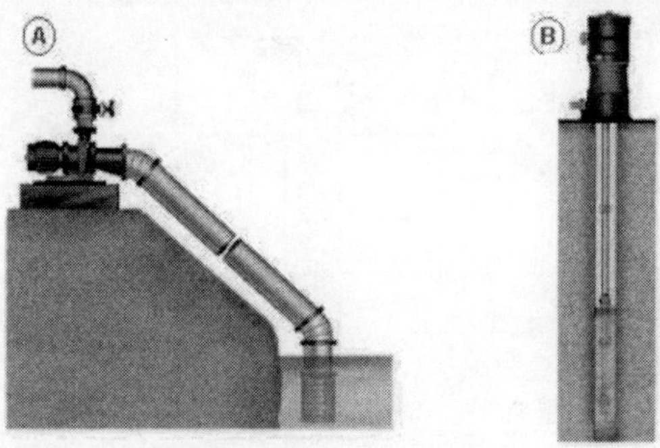

FIGURE 3.47 - Left: a rotodynamic pump powered by a tractor. Right: a pumping unit with rotodynamic pumps in parallel.

Air-lift pumps.A rising main, which is submerged in a well so that more of it is below the water level than above it, has compressed air blown into it at its lowest point. The compressed air produces a froth of air and water, which has a lower density than water and consequently rises to the surface. The compressed air is usually produced by an engine driven air compressor (Figure 3.48). The main advantage of the air-lift pump is that there are no mechanical below-ground components to wear out, so it is essentially simple, reliable, virtually maintenance-free and can easily handle sandy or gritty water. The disadvantages are rather severe: first it is inefficient as a pump, probably no better than 20-30 percent in terms of compressed air energy to hydraulic output energy and this is this is compounded by the fact that air compressors are also general inefficient. Second, it usually requires a borehole to be drilled considerably deeper (more than twice the depth of the static water level) than otherwise would be necessary.

FIGURE 3.48 - Scheme of an air-lift pump.

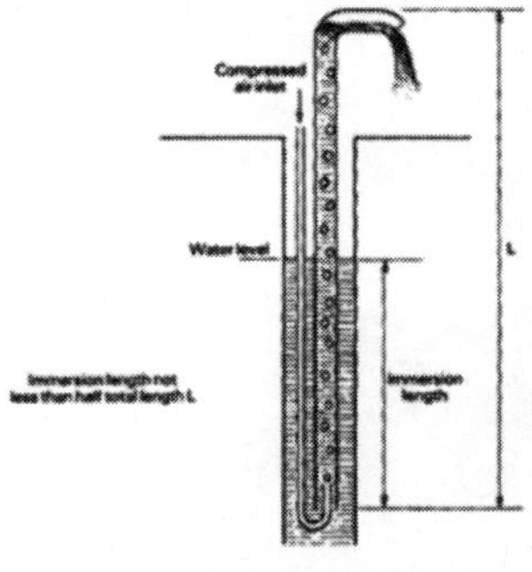

Impulse (water hammer) pumps. These devices apply the energy of falling water to lift a fraction of the flow to a higher level than the source. The principle they work by is to let the water from the source flow down a pipe and then to create sudden pressure rises by intermittently letting a valve in the pipe slam shut. This causes a "water hammer" effect which results in a sudden sharp rise in water pressure sufficient to carry a small proportion of the supply to a considerably higher level. They therefore are applicable mainly in hilly regions in situations where there is a stream or river flowing quite steeply down a valley floor, and areas that could be irrigated which are above that level that can be commanded by small channels contoured to provide a gravity supply. A practical example of this pump is the hydraulic ram pump or "hydram" (Figure 3.49). The main virtue of the hydram is that it has no substantial moving parts, and is therefore mechanically extremely simple, which results in very high reliability, minimal maintenance requirements and a long operational life. However, in most cases the output is rather small (in the region of 1–3 litre/sec) and they are therefore best suited for irrigating small-holdings.

FIGURE 3.49 - Schematic diagram of hydram installation.

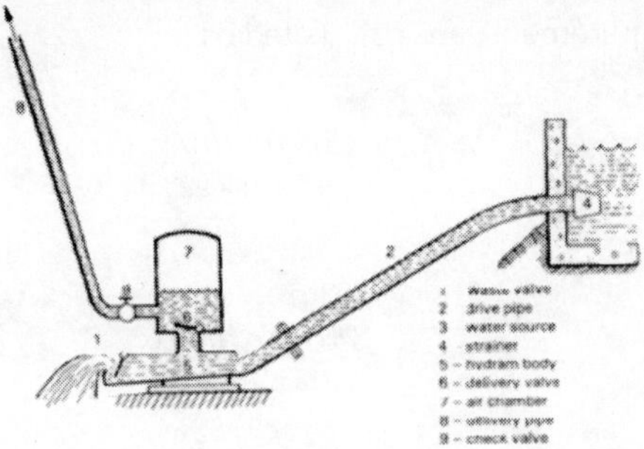

Gravity devices. Syphons are the most common device of this type, though strictly speaking they they are not water-lifting devices, since, after flowing through a syphon, water finishes at a lower level than it started. However syphons can lift water over obstructions at a higher level than the source and they are therefore potentially useful in irrigation (Figure 3.50). Syphons are limited to lifts about 5 m at sea level for exactly the same reasons related to suction lifts for pumps. The main problem with syphons is that due to the low pressure at the uppermost point, air can come out of solution and form a bubble, which initially causes an obstruction and reduces the flow of water, and which can grow sufficiently to form an airlock which stops the flow. Therefore, the syphon pipe, which is entirely at a subatmospheric pressure, must be completely air-tight.

FIGURE 3.50 - A syphon arrangement.

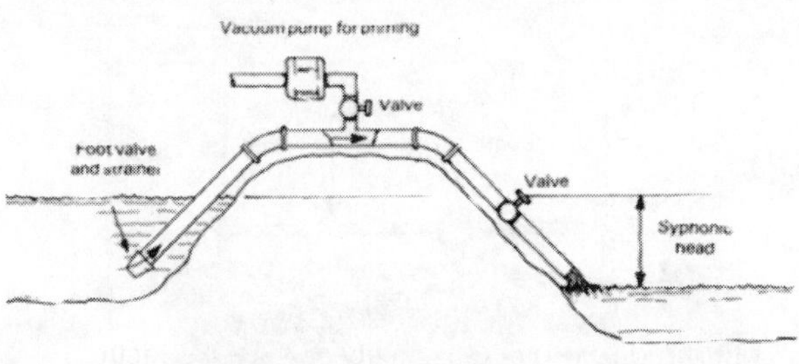

Calculation of the power requirements (P)?

The power requirements are calculated as:

$$P(HP) = \frac{Q(l/s) \times Ht(m)}{75 \times e1 \times e2}$$

and

$$P(kw) = \frac{Q(l/s) \times Ht(m)}{102 \times e1 \times e2}$$

where:
Ht is the total head;
e1 is the pump efficiency (fraction in the order of 0.5–0.8); and
e2 is the driving efficiency (fraction of 0.7–0.9 for electric motors and 0.5–0.75 for diesel engines).

The total head (Ht) required for the normal operation of the system is the sum of the following pressures (Figure 3.51):

FIGURE 3.51 - Ht is the total head, Ha is the elevation head, Hi is the emitter operation head and Hp is the friction losses head (sum of friction losses in the main line, submains, manifolds, laterals, valves, pipe fittings and minor losses).

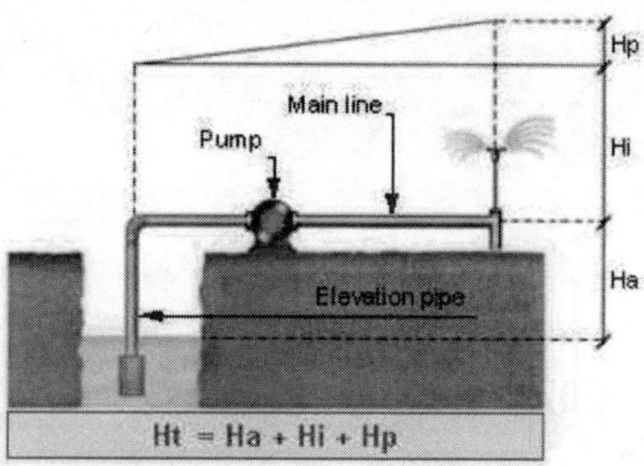

CHAPTER 4: System design

INTRODUCTION

The engineering design is the second stage in irrigation planning. The first stage is the consideration of the crop water requirements, the type of soil, the climate, the water quality and the irrigation scheduling. The water supply conditions, the availability of electricity and the field topography also need to be considered. The economic considerations, the labour and the know-how also need to be taken into account. The irrigation system is selected after a thorough evaluation of the above data and the computation of the system's flow, the irrigation dose, the duration of application and the irrigation interval.

Once the design has been completed, a detailed list of all the equipment needed for the installation of the system must be prepared with full descriptions, standards and specifications for every item.

SYSTEM DESIGN

The engineering and hydraulic design procedure is almost the same in all kinds of pressurized irrigation systems. It consists of a series of interlinked calculations. The various stages are outline below.

Selection of the water emitter (sprinkler, dripper, minisprinkler, bubbler, hose, etc.) according to the crop, irrigation method and requirements:

- type, flow rate, operating pressure, diameter coverage;
- spacing and number per lateral line.

DESIGN OF THE LATERALS

- length, direction, spacing and total number of lateral lines (in solid systems) or lateral positions (in semi-permanent installations);
- flow of the lateral = number of emitters per lateral *x* emitter flow rate;
- number of laterals operating simultaneously = system flow/flow of lateral;
- number of shifts to complete one irrigation = total number of lateral lines or positions ÷ number of laterals operating simultaneously;

- Duration of application = irrigation dose in millimetres ÷ application rate in millimetres per hour, or irrigation dose in cubic metres ÷ system flow in cubic metres per hour.

DETERMINATION OF THE SIZE OF THE PIPELINES

Lateral lines

It is important to understand the water emitter's functions and principle of operation before commencing the design process. One of the main characteristics of all types of emitters is the relationship between flow rate and operating pressure, which is usually expressed by the empirical formula:

$$q = kdH\ *$$

where ***q*** is the emitter discharge, ***k*** and ***d*** are coefficients (constants), ***H*** is the pressure at the emitter and * is an exponent characterized by the emitter flow regime and the flow rate curve as a function of the pressure.

The lower the value of *, the less the influence of pressure variations on the emitter flow rate along the lateral line. Most of the water emitter flow regime is fully turbulent with an exponent value equal to 0.5. Thus, the difference in discharge is half the difference in pressure, when the ratio of the two different pressures is $< 1.3/1.0$.

In order to ensure a high uniformity of water application over the field, the differences in the discharge of the emitters should be kept to the minimum possible and in no case exceed 10 percent. These criteria were established by J. Christiansen for sprinklers and are now applied in all pressurized systems. As a general rule, the maximum permissible difference in pressure between any two emitters in operation should be no more than 20 percent. The lateral lines with emitters must be of a size that does not allow a loss of head (pressure) due to friction of more than 20 percent.

The loss of head due to friction (friction losses) in lateral pipes is taken from a graph or a table. The reading is usually given as loss of head of water in metres or feet per 100 m or 100 ft of pipe. For example, in a 50 mm quick coupling sprinkler lateral pipe with a 15 m^3/h flow, the friction losses are 7 percent. If the length of the lateral is 120 m, the friction losses are: 7/100 120 = 8.4 m. However, this figure is for the total flow of 15 m^3/h running the whole length of lateral. Thus, it is not the true figure as the flow is distributed en route through the emitters. In order to compute the actual losses the above figure is multiplied by Christiansen's reduction coefficient, F, to compensate for the water delivered along the lateral line. The F values depend on the number of the outlets uniformly spaced along the pipeline (Table 4.1).

Three different series of F values exist corresponding to the Q exponent (m) of the three main friction loss formulas: Hazen Williams, 1.85; Scobey, 1.9; and Darcy Weisbach, 2.0. Moreover, lower values are taken if the distance of the first outlet is half the spacing of the outlets, etc. However, the differences between the various F values are almost negligible.

TABLE 4.1 - F factor for multiple outlets

Number of outlets	F value (m = 2.0)	Number of outlets	F value (m = 2.0)
1	1.0	12	0.376
2	0.62	15	0.367
3	0.52	20	0.360
4	0.47	24	0.355
5	0.44	28	0.351
6	0.42	30	0.350
7	0.41	40	0.345
8	0.40	50	0.343
9	0.39	100	0.338
10	0.385	>	0.333

Assuming that in the above example there are ten emitters (in this case sprinklers) on the lateral, the F value is 0.4. Then, in a 50 mm quick coupling lateral, 120 m long, with a flow of 15 m³/h, with 10 sprinklers of 1.5 m³/h at 2.0 bars, the friction losses are: 7/100 x 120 x 0.4 = 3.36 m head of water. This figure must not exceed the maximum permissible, which is 20 percent of the emitter's average operating pressure, i.e. 2.0 bars x 0.20 = 0.4 bars (4 m) on level ground. Where the lateral slopes downwards, the difference in elevation is added to the maximum permissible loss of pressure. Similarly, it is deducted where the lateral slopes upwards.

Due to the multiplicity of emitters with variable flow regimes and other factors affecting the pressure/discharge relation along the laterals in the field, such as local minor losses that occur at the connection of the emitters on small-sized pipes and temperature fluctuations, the manufacturers should always provide charts for the optimum length of emitter laterals, based on the size of pipe, emitter spacing, operating pressure, flow rate and slope.

Manifolds, submain and main pipelines

On the manifolds, whether these pipelines are the submains or the mains as well, a number of laterals are fed simultaneously. The flow of the line is distributed en route, as in the laterals with the emitters. Consequently, when computing the friction losses, Christiansen's reduction coefficient, F, is also considered. Example: 120 m of 75 mm HDPE, 6 bars, manifold line, 16.3 m^3/h, 6 laterals operating simultaneously; the friction loss under full flow is 3.3 percent, i.e. 4.0 m x 0.42 = 1.7 m approximately.

The mains, submains and all hydrants are selected in such sizes that the friction losses do not exceed approximately 15 percent of the total dynamic head required at the beginning of the system's piped network. On level ground, these friction losses amount to about 20 percent of the emitter's fixed operating pressure. This is a practical rule for all pressurized systems to achieve uniform pressure conditions and water distribution at any point of the systems. Figure 4.1 below should not be confused with or related in any way to the maximum permissible friction losses along the laterals.

FIGURE 4.1 - Operating pressure required for pressurized systems.

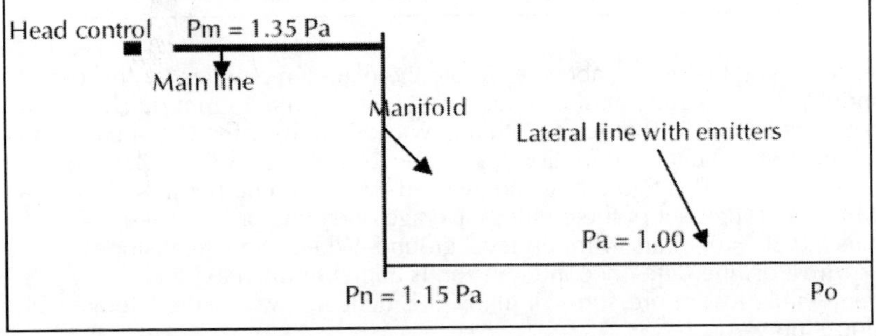

In the above figure, Pa is the average emitter pressure, or fixed pressure taken from the catalogue; Pn is the lateral inlet pressure; Po = 0.95 Pa is the distal end emitter pressure; and Pm is the pressure at the inlet of the main line.

Pn - Po = 0.20 Pa;
Po = Pn ÷ 1.21;
Pn = 1.15 Pa;
Pm = 1.35 Pa.

The friction loss in a lateral with emitters is very high at the beginning and drops rapidly after the first few outlets and then more gradually toward the end of the line. In the upper one-fourth of the lateral the friction loss is approximately 75 percent of the total. Another important element is the flow velocity in the mains, submains and hydrants. This value should always be kept below 1.7 m/s in plastic tubes and a maximum of 2 m/s in other pipes (steel, aluminium, etc.). From the flow velocity formula, V = Q/A, the pipe inside diameter is determined for a given flow:

$$diameter(mm) = \sqrt{\frac{Qm^3/hr}{V(m/s)}} \times 18.8$$

HEAD CONTROL

The component parts of the head control and their size are in accordance with the system requirements. In micro-irrigation systems the units are complete with filters and fertilizer injectors, while in sprinkler and hose irrigation systems the head controls are simple with the minimum of equipment. The friction losses in the various component parts vary accordingly from 3 to 10 m.

The friction loss formulas are empirical and include many variables and correction factors. In calculating the pipe friction losses from equations, extensive practical experience is needed. In view of the fact that great accuracy is not possible due to the unpredictable changes in pipe roughness, water viscosity, nozzle wear, clogging, etc., the use of friction loss tables and nomographs is recommended.

TOTAL DYNAMIC HEAD OF THE SYSTEM

The total pressure head or dynamic head required for the normal operation of the system is the sum of the following pressures (Table 4.2):

TABLE 4.2 - Total pressure head of system

Pressure at the emitter	Metres head
Friction losses in the lateral line	Metres head
Friction losses in the manifold	Metres head
Friction losses in the submains and in the main line	Metres head
Friction losses in the valves and pipe fittings and minor losses (usually up to 15 percent of the total losses in the pipes)	Metres head
Difference in elevation (plus or minus)	Metres head
Loss of pressure in the head control	Metres head
Total pressure head of system	**Metres head**

TOTAL DYNAMIC HEAD OF THE PUMPING UNIT

This is the sum of the system's total head plus the pumping lift. The brake horsepower formula is:

$$BHP = \frac{Q \times TDH}{270 \times e1 \times e2}$$

where ***Q*** is the flow capacity in cubic metres per hour, ***TDH***is expressed in metres, ***e1*** is the pump efficiency (fraction), **e2** is the driving efficiency (fraction), and 270 is a constant for metric units.

- Pump efficiency: 0.5–0.8;
- Electric motor efficiency: 0.7–0.9;
- Diesel engine efficiency: 0.5–0.75.

The overall pumping efficiency under field conditions ranges accordingly from 0.35 in engine driven units to 0.50 in motor driven pumps. Higher efficiencies are not realistic.

CHAPTER 5: Equipment, standards and tenders for supply

With the completion of the design, a detailed list of all the equipment needed (bill of quantities) for the installation of the system must be prepared with full descriptions, standards and specifications for every item. The preparation of this list is of great importance. In addition to the quantities, it is imperative to determine and specify:

- size and name (2 inches ball valve, 50 mm pipe, etc.);
- kind of material (brass, uPVC, etc.);
- pressure rating (PN 16 bars, 6 bars, etc.);
- type of joints (screw, solvent welded, etc.);
- standards complied with (ISO 161, 3606, BS 21, ISO 7, etc.).

Three different lists may be prepared: one for the mains, submains and manifolds with the hydrants; one for the laterals with the emitters; and one for the head control. Sizes will have already been decided on during the design stage.

WORKING PRESSURE OF THE EQUIPMENT

A closed pipe pressurized system installation consists of pipes of different working pressures according to the location. The main pipelines are subjected to higher pressures than the submains, manifolds and laterals, therefore the main lines should be stronger than the other pipelines. The working pressure of the pipes to be installed should always be higher than the system's operating pressure. For example, in a micro-jet (minisprinkler) installation the approximate operating pressure is 2.3–2.5 bars in the laterals, 2.5–2.7 bars in the manifolds and 2.7–3.0 bars in the mains. A pipe working pressure of 4.0 bars seems to meet the requirements of the system. However, although the low to medium pressure systems are not subjected to the very high pressures created by water hammer, it is advisable to use 6.0 bar pipes for the main line and 4.0 bars for the other pipelines.

MAIN, SUBMAIN AND MANIFOLD PIPELINES, AND HYDRANTS

The most widely used kinds of pipes for these lines are rigid PVC, HDPE, LDPE and quick coupling light steel or aluminium. The following must be determined:

- total length and pieces of the pipelines (about 5 percent should be added to the total);
- quantities of pipe connector fittings (bends, tees, end plugs, reducers, etc.) of the same type of connection to be used with the above pipes;
- number of bends, tee outlets and clamp saddles, which have two different types of connections, e.g. tee 90 mm *x* 90 mm *x* 2 inches (internal threaded), bend 110 mm *x* 3 inches (flanged);
- quantity of adaptors (starters). These fittings have one end threaded or flanged and the other end arranged in the same type of connection as the pipes. They are used at the starting point of the pipelines and at any other point where valves are fitted;
- number of shut-off and air valves on the distribution network. The air valves are fitted on riser pipes connected with clamp saddles on the mains;
- quantities of the riser pipes for the hydrants, if the mains are buried, and of the shut-off valves or the special hydrant valves. If the mains are not buried, then the fittings for connecting clamp saddles with the shut-off valves must be determined. The number of these fittings is equal to the number of the hydrants.

LATERALS

Quick coupling and LDPE pipes are used as surface laterals in the majority of the systems. The following must be determined:

- total length of pipes required;
- quantities of adaptors, tees, bends, end plugs and line filters;
- total number of emitters and their connector fittings if any, e.g. minisprinkler complete with plunger, connecting flexible tube and plastic wedge, or specify in terms of set, e.g. minisprinkler complete set.

HEAD CONTROL

All the components of the head control of the system must be determined, i.e. shut-off valves, check valve, air valve, fertilizer injector, filters, pressure regulators, etc. In addition, all the auxiliary fittings must be included, such as the pipe pieces, hoses and fittings needed to assemble the unit, and the pressure gauges and other small devices required.

PUMPING UNIT

A full and detailed description of the pumping unit must be given, including the following:

- the average break horsepower (BHP) calculated of the driving force and the type (engine or motor);
- the kind of pump (centrifugal single or multi-stage, turbine, electrosubmersible), the inlet and outlet diameter, and the type and number of stages;
- the capacity and output of the pumping unit, i.e. the water delivery versus the dynamic head.

STANDARDS

Standards are consensus documents developed by task groups and technical committees to specify mechanical, functional and other requirements of irrigation equipment (workmanship, materials, dimensions, pressures, test methods, supply and delivery conditions). The continuing development of new standards follows the accelerated pace of the technological development in the field of pressurized irrigation techniques. However, some old standards may be used as an indication of the basic needs. All pipes, pipe fittings and other irrigation equipment are manufactured according to various standards applied in the countries of origin. These standards, although equivalent to each other, vary in terms of the dimensioning, the class rating, the safety factor and the nomenclature. Much technical engineering effort has been devoted by the International Standards Organization (ISO) to the establishing of international standards and specifications so that all national and regional standards are in broad conformity.

The general policy in many countries is to adopt gradually the European Standards (EN) and to withdraw any conflicting national standards. The EN are prepared by the Technical Committees of the European Committee for Standardization (CEN/TC) and is given the status of national standards in most for the European countries and the United Kingdom. Each EN forms part of a System Standard based on the results of the work undertaken by corresponding ISO/TC. The System Standards are consistent with general standards on function and installation and supported by separate standards on test methods to which references are made throughout the System Standards. Also EN incorporate provisions of other publications.

However, at present the variety of standards causes small farmers a great deal of confusion regarding thermoplastic irrigation equipment. In Table 5.1 below is an example of a 4 inches rigid PVC pipe, 6.0 bars, in two different national standards:

TABLE 5.1 - A 4 inches rigid PVC pipe (6.0 bars) in two different national standards

	to DIN 8062	**to ASTM D2241 (SDR 4.1)**
Nominal diameter	110.0 mm	4 inches
Outside diameter	110.0 mm	114.3 mm
Inside diameter	103.6 mm	108.7 mm
Wall thickness	3.2 mm	2.8 mm
Working pressure	6.0 bars	6.8 bars (100 psi)

The description of the equipment should be as clear and simple as possible. An example with the minimum specifications required for two items is as follows:

Item 1: Black LDPE pipe, PN 4.0 bars, to DIN 8072 or equivalent standards in compliance with ISO standards, supplied in coils of 200 m:
a. 32 mm DN, 1 800 m;
b. 25 mm DN, 3 200 m.

Item 2: Polypropylene connector fittings manufactured to ISO metric dimensions. Quick release, compression type and/or threaded (screw-type) ends male or female, to ISO 7 or BS 21, PN 10 bars for use with the above PE pipes:
a. 63 mm x 2 inches (male) adaptor, 7 pieces;
b. 63 mm x 2 inches (female) clamp saddle, 2 pieces;
c. 50 mm x 2 (male) adaptor, 2 pieces.

Should the equipment not comply with any standard, due to many reasons, a full technical description should be given of the material it is made of, the working pressure and the use. The latter is important because the fittings should be made of material recommended for use with the particular pipe.

Most of the irrigation equipment should meet the appropriate material, dimensional, and quality requirements recommended in the specifications in Table 5.2.

TABLE 5.2 - Equipment standards and specifications

Standard name:	Standard description:
ASAE EP419.1 February 2003	Evaluation of Irrigation Furrows
ASAE EP405.1 february 2003	Design and Installation of Microirrigation Systems
ANSI/ASAE S261.7 December 2001	Design and Installation of Nonreinforced Concrete Irrigation Pipe Systems
ASAE S526.2 January 2001	Soil and Water Terminology
ASAE S491 February 2003	Graphic Symbols for Pressurized Irrigation System Design
ANSI/ASAE S395 February 2003	Safety for Self-Propelled, Hose-Drag Agricultural Irrigation Systems
BSR/ASAE S577-200x	Specification for Poly (Vinyl Chloride) (PVC) Irrigation Pipe (PIP) Fittings
ANSI/ASAE S376.2 February 2004	Design, installation and Performance of Underground, Thermoplastic Irrigation Pipelines
ANSI/ASAE S436.1 December 2001	Test Procedure for Determining the Uniformity of Water Distribution of Center Pivot and Lateral Move Irrigation Machines Equipped with Spray or Sprinkler Nozzles
ANSI/ASAE S330.1 February 2003	Procedure for Sprinkler Distribution Testing for Research Purposes
ANSI/ASAE S539 February 2003	Media Filters for Irrigation - Testing and Performance Reporting
ASAE S447 February 2003	Procedure For Testing and Reporting Pressure Losses in Irrigation Valves
ANSI/ASAE S397.2 February 2003	Electrical Service and Equipment for Irrigation
ASAE EP409.1 February 2004	Safety Devices for Chemigation
ASAE S435 February 2004	Polyethylene Pipe Used for Microirrigation Laterals
ASAE S398.1 january 2001	Procedure for Sprinkler Testing and Performance Reporting
ASAE EP367.2 February 2003	Guide for Preparing Field Sprayer Calibration Procedures
ASAE S327.2 February 2003	Terminology and Definitions for Agricultural Chemical Application
ANSI/ASAE S553 march 2001	Collapsible Emitting Hose (Drip Tape) – Specifications and Performance Testing
ASAE EP369.1 December 1999	Design of Agricultural Drainage Pumping Plants
ASAE S561 February 2003	Procedure for Measuring Drift Deposits from Ground, Orchard, and Aerial Sprayers
ASAE EP400.2T February 2003	Designing and Constructing Irrigation Wells
ASAE EP285.7 January 2001	Use of SI (Metric) Units
ASAE S431.3 February 1999	Safety Signs
ANSI/ASAE S397.2 February 2003	Electrical Service and Equipment for Irrigation
ASAE S471 February 2003	Procedure for Measuring Sprayer Nozzle Wear Rate
ASAE S263	Minimum Standards for Aluminum Tubing
ISO 7714:2000	Agricultural irrigation equipment – Volumetric valves – General requirements and test methods
ISO 7749-1:1995	Agricultural irrigation equipment – Rotating sprinklers – Part 1: Design and operational requirements
ISO 8026:1995	Agricultural irrigation equipment – Sprayers –
ISO 8026:1995/Amd 1:2000	General requirements and test methods
ISO/TR 8059:1986	Irrigation equipment – Automatic irrigation systems – Hydraulic control
ISO 8224-1:2003	Traveller irrigation machines – Part 1: Operational characteristics and laboratory and field test methods
ISO 8224-2: 1991	Traveller irrigation machines – Part 2: Soft wall hose and couplings – Test methods
ISO 8779:2001	Polyethylene (PE) pipes for irrigation laterals – Specifications
ISO 8796:2004	Polyethylene (PE) 32 and PE 40 pipes for irrigation laterals – Susceptibility to environmental stress-cracking induced by insert-type fittings – Test method and requirements
ISO 9261:2004	Agricultural irrigation equipment – Emitters and emitting pipe – Specification and test methods
ISO 9625:1993	Mechanical joint fittings for use with polyethylene pressure pipes for irrigation purposes

TABLE 5.2 - Equipment standards and specifications (cont'd)

Standard name:	Standard description:
ISO 9635-1:2006	Agricultural irrigation equipment – **Irrigation valves – Part 1: General requirements**
ISO 9635-2:2006	Agricultural irrigation equipment – **Irrigation valves – Part 2: Isolating valves**
ISO 9635-3:2006	Agricultural irrigation equipment – **Irrigation valves – Part 3: Check valves**
ISO 9635-4:2006	Agricultural irrigation equipment – **Irrigation valves – Part 4: Air valves**
ISO 9635-5:2006	Agricultural irrigation equipment – **Irrigation valves – Part 5: Control valves**
ISO 9644:1993	Agricultural irrigation equipment – Pressure losses in irrigation valves –
ISO 9644:1993/Amd 1:1998	Test method
ISO 9911:2006	Agricultural irrigation equipment – Manually operated small plastics valves
ISO 9912-1:2004	Agricultural irrigation equipment – **Filters for micro-irrigation – Part 1: Terms, definitions and classification**
ISO 9912-2:1992	Agricultural irrigation equipment – **Filters – Part 2: Strainer-type filters**
ISO 9912-3:1992	Agricultural irrigation equipment – **Filters – Part 3: Automatic self-cleaning strainer-type filters**
ISO 10522:1993	Agricultural irrigation equipment – Direct-acting pressure-regulating valves
ISO 11545:2001	Agricultural irrigation equipment – Centre-pivot and moving lateral irrigation machines with sprayer or sprinkler nozzles – Determination of uniformity of water distribution
ISO 11678:1996	Agricultural irrigation equipment – Aluminium irrigation tubes
ISO 11738:2000	Agricultural irrigation equipment – Control heads
ISO 12347:1995	Agricultural irrigation – Wiring and equipment for electrically driven or controlled irrigation machines
ISO 13457:2000	Agricultural irrigation equipment – Water-driven chemical injector pumps
ISO 13460:1998	Agricultural irrigation equipment – Plastics saddles for polyethylene pressure pipes
ISO 15081:2005	Agricultural irrigation equipment – **Graphical symbols for pressurized irrigation systems**
ISO 15873:2002	Irrigation equipment – Differential pressure Venturi-type liquid additive injectors
ISO 15886-1:2004	Agricultural irrigation equipment – **Sprinklers – Part 1: Definitions of terms and classification**
ISO 15886-3:2004	Agricultural irrigation equipment – **Sprinklers – Part 3:Characterization of distribution and test methods**
ISO 16149:2006	Agricultural irrigation equipment – **PVC above-ground low-pressure pipe for surface irrigation – Specifications and test methods**
ISO 4065	Thermoplastics pipes – Universal wall thickness table
ISO 7-1:1994	Pipe threads where pressure-tight joints are made on the threads – **Part 1: Dimensions, tolerances and designation**
ISO 7-2:2000	Pipe threads where pressure-tight joints are made on the threads – **Part 2: Verification by means of limit gauges**
ISO 49:1994/Cor 1:1997	Malleable cast iron fittings threaded to ISO 7-1
ISO 4422-1:1996	Pipes and fittings made of unplasticised poly(vinyl chloride) (PVC-U) for water supply – Specifications – Part 1: General
ISO 4422-2:1996	Pipes and fittings made of unplasticised poly(vinyl chloride) (PVC-U) for water supply – Specifications - Part 2: Pipes (with or without integral sockets)
ISO 4422-3:1996	Pipes and fittings made of unplasticised poly(vinyl chloride) (PVC-U)
ISO 4422-4:1997	Pipes and fittings made of unplasticised poly(vinyl chloride) (PVC-U) for water supply – Specifications - Part 4: Valves and ancillary equipment
ISO 4422-5:1997	Pipes and fittings made of unplasticised poly(vinyl chloride) (PVC-U) for water supply – Specifications - Part 5: Fitness for purpose of the system for water supply – Specifications - Part 3: Fittings and joints
ASTM D1785-06	Standard Specification for Poly(Vinyl Chloride) (PVC) Plastic Pipe, Schedules 40, 80, and 120

TABLE 5.2 - Equipment standards and specifications (cont'd)

Standard name:	Standard description:
ASTM D2241-05	Standard Specification for Poly(Vinyl Chloride) (PVC) Pressure-Rated Pipe (SDR Series)
ASTM D2447-03	Standard Specification for Polyethylene (PE) Plastic Pipe, Schedules 40 and 80, Based on Outside Diameter
ASTM D2464-99	Standard Specification for Threaded Poly (Vinyl Chloride) (PVC) Plastic Pipe Fittings, Schedule 80
ASTM D2466-02	Standard Specification for (Vinyl Chloride) (PVC) Plastic Pipe Fittings, Schedule 40
ASTM D2467-02	Standard Specification for (Vinyl Chloride) (PVC) Plastic Pipe Fittings, Schedule 80
ASTM D2609-02	Standard Specification for Plastic Insert Fittings for Polyethylene (PE) Plastic Pipe
ASTM D2683-98	Standard Specification for Socket-Type Polyethylene Fittings for Outside Diameter-Controlled Polyethylene Pipe and Tubing
ASTM D2683-04	Standard Specification for Socket-Type Polyethylene Fittings for Outside Diameter-Controlled Polyethylene Pipe and Tubing
ASTM D3139-98(2005)	Standard Specification for joints for Plastic Pressure Pipes Using Flexible Elastometric Seals
ASTM D3261-03	Standard Specification for Butt Heat Fusion Polyethylene (PE) Plastic Fittings for Polyethylene (PE) Plastic Pipe and Tubing
BS 21:1985	Specification for pipe threads for tubes and fittings where pressure-tight joints are made on the threads (metric dimensions) (equivalent to ISO 7-2:1982)
BS 3867:1987	Method of specifying outside diameters and pressure ratings for pipe of thermoplastics materials (inch series) (equivalent to ISO 161-2:1977)
BS 4346 (Part 1–3)	Joints and fittings for use with unplasticized PVC pressure pipes
BS 143 and 1256:2000	Threaded pipe fittings in malleable cast iron and cast copper alloy
DIN 2440/41/42	Steel tubes (Medium-Weight) suitable for screwing
DIN 2999 (1–6)	Pipe threads for tubes and fittings
DIN 8062 (1988)	Unplasticised polyvinyl chloride (PVC-U, PVC-HI) pipes – Dimensions
DIN 8072 (1987)	Pipes of low-density PE (low-density polyethylene) – Dimensions
DIN 8074 (1999)	High-density polyethylene (PE-HD) pipes – Dimensions
DIN 8075 (1999)	High-density polyethylene (PE-HD) pipes – Testing
DIN 8161 (1994)	Unplasticised polyvinyl chloride pipes – General quality requirements and testing
EN 2452-2	Plastic piping systems for water supply - Unplasticized Poly(vinil chloride) (PVC-U) – Part 2: Pipes
EN 12201-2	Plastic piping system for water supply – Polyethylene (PE) – Part 2: Pipes

Note:

ASAE: The Society for Engineering in Agriculture, Food, and Biological Systems (former American Society of Agricultural Engineers).
ANSI: American National Standards Institute.
ASTM: American Society for Testing Material.
BS: British Standards.
DIN: Deutsches Institut für Normung (German standards).
ISO: International Standards Organization.
EN: European Standard.

TENDERS

The purchasing of irrigation equipment or execution of services, such as the installation, operation and maintenance of irrigation networks and or pumps, should be subject to public tender.

For equipment and services up to a value of US$500, the purchase can be effected through 'quotations', i.e. written quotations may be asked from a representative number (2–3) of suppliers. Where the value of the equipment exceeds a certain amount, e.g. US$600, their purchase should be affected through tender. This is done in accordance with the 'stores regulations' applied in the project or the country concerned.

Wide publicity should be given to every 'notice inviting tenders' (invitation for tenders). This must include the name of the buyer, a brief description of the items for which tenders are invited, the address for delivery of equipment, and the closing date and time of the tenders. Moreover, it should include a statement that the buyer is not bound to accept the lowest or any other tender, and also state to whom the bidders must apply for full particulars.

In the case of 'local tenders' for the purchase of relatively limited quantities, the tender document that must be available and given to prospective bidders on request should include only the general conditions of the tender and the technical specifications of goods. It is important that all required conditions be clearly stated in detail in the tender document, including the time and method of delivery, i.e. FOB (Free on board), CIF (Cost insurance and freight), ex-stock; method of payment, i.e. letter of credit, cash against documents, payment on delivery, etc.; and other related information. For tenders over US$3 000, bidders should furnish a bank guarantee or cheque equal to 10 percent of the value of the tender price. An example of this kind of tender is given below

In the case of 'international bids', the contract documents must include, in detail, the following:

- invitation for bids (as described above);
- instructions to bidders (source of funds, eligible bidders, goods and services, cost, content of bidding documents, preparation and submission of bids, opening and evaluation, award of contract, etc.);
- general conditions of contract (definitions, country of origin and standards, performance, security, inspection and tests, insurance, transportation, warranty, payment, amendments, delays, force majeure, etc.);
- special conditions;
- technical specifications (general, materials and workmanship, schedules of requirements/bill of quantities [Table 5.3], and particular technical requirements/specifications [Table 5.4]);
- bid form and price schedules;
- contract form, bid security and performance security.

EXAMPLE

Tenders for the supply of irrigation equipment

Tenders are hereby invited for the supply of irrigation equipment required for a private farm in the Project area, as per attached quantities, description and specification.

General conditions of tenders

1 **Price:** Bidders to quote prices per unit and total, CIF nearest port, Republic of ..., full liner terms, including bank charges on the attached price schedules. Prices to be firm for at least 90 calendar days from the closing date of tender.

2 **Delivery:** Date of delivery in the project site should not exceed 60 days from the time of awarding the tender.

3 The tenders should be sealed and addressed to the General Manager, Irrigation Project, P.O. Box 5564. Tenders should be marked 'TENDER FOR THE SUPPLY OF IRRIGATION EQUIPMENT FOR PRIVATE FARM' on the envelope and should reach Project main offices not later than 31 December 2007.

4 The bidder shall be prepared to accept the prices tendered by him. The tender shall become binding and be carried into effect upon being accepted by the Project. Should the bidder delay execution of the tender or refuse to execute the tender, the bidder shall be liable for any expenses incurred by the Project.

5 **Payment:** The Project shall make all necessary arrangements towards the opening of the letter of credit in US dollars for goods to be supplied in its name and on behalf of the supplier within seven days after receiving the import licence. The Project shall make a first payment of 50 percent of the value of the tender upon submission of all the necessary shipping documents. Such documents shall reach the Project at least one month before the scheduled date of arrival of goods into the port of entry. A second payment of 50 percent of the value of contract shall be paid to the supplier after receipt of goods at Project's store and issuing certificate of acceptance in accordance with the technical specifications.

6 Insurance to cover all the risks for the CIF value, plus 10 percent from warehouse up to the Project's stores.

7 Bidders to quote the country of origin. It is imperative to quote for the items according to the specifications and standards as per the attached list. Otherwise, full details are required.

8 Bidders should provide a guarantee of excellent workmanship and against faulty material of not less than 12 months.

9 Selected candidates should confirm by fax without delay their receipt of the invitation to tender.

10 Tenders shall not be considered unless all the above conditions have been strictly observed.

11 Tenders to be submitted in duplicate.

12 The Project does not bind itself to accept the lowest or any tender.

TABLE 5.3 - Bill of quantities

Item	Description	Unit	Quantity	Rate (US$)	Amount (US$)
1.	HDPE pipe ø 75 mm	m	300		
2.	HDPE pipe ø 63 mm	m	650		
3.	HDPE pipe ø 50 mm	m	100		
4.	LDPE pipe ø 25 mm	m	3 600		
5.	LDPE pipe ø 16 mm	m	1 400		
6.	Clamp saddle ø 75 mm x 2 in (F)	pcs	8		
7.	Clamp saddle ø 63 mm x ¾ in (F)	pcs	70		
8.	Clamp saddle ø 50 mm x ¾ in (F)	pcs	10		
9.	Adaptor (starter) ø 75 mm x 3 in (M)	pcs	1		
10.	Adaptor (starter) ø 63 mm x 2 in (M)	pcs	7		
11.	Adaptor (starter) ø 50 mm x 2 in (M)	pcs	1		
12.	Adaptor (starter) ø 25 mm x ¾ in (M)	pcs	240		
13.	Adaptor (starter) ø 16 mm x ¾ in (M)	pcs	150		
14.	Coupling ø 75 mm	pcs	2		
15.	Coupling ø 63 mm	pcs	4		
16.	Coupling ø 50 mm	pcs	1		
17.	Coupling ø 25 mm	pcs	30		
18.	Coupling ø 16 mm	pcs	10		
19.	Tee ø 50 x 50 x 50 mm	pcs	1		
20.	Tee ø 25 x 25 x 25 mm	pcs	10		
21.	Tee ø 25 mm x ¾ in (M)	pcs	10		
22.	Tee ø 25 mm x ½ in (F)	pcs	150		
23.	Cross ø 2"	pcs	1		
24.	Nipple hexagon ø 2 in	pcs	8		
25.	Nipple hexagon ø ¾ in	pcs	80		
26.	End plug ø 75 mm	pcs	1		
27.	End plug ø 63 mm	pcs	10		
28.	End plug ø 50 mm	pcs	1		
29.	Ball valve ø 2 in	pcs	8		
30.	Ball valve ø ¾ in	pcs	80		
31.	Filter strainer ø 3 in	pcs	1		
32.	Dripper emitter 24 litres/h	pcs	5 000		
33.	Sprinkler pop-up full circle	pcs	4		
34.	Air valve ¾ in	pcs	2		
35.	Valve box	pcs	8		
36.	Excavation of trench and backfill	m	1 050		

TABLE 5.4 - Equipment specification

Item number	Equipment specification
1,2,3	Black HDPE pipes, PN 6.0 bars, in accordance with CYS104: Part 1: 1985 (Cyprus Standard) or equivalent other national standards in compliance with ISO. Supplied in 100 and 60 m rolls.
4,5	Black LDPE pipes, PN 4.0 bars, in accordance with CYS106: Part 1, Part 2: 1985 or equivalent other national standards in compliance with ISO. Supplied in 200 m rolls.
6-28	Polypropylene connector fittings for use with PE Pipes to CYS and ISO dimensions, quick release compression type and/or screw ends to BS 21, or ISO 7, PN 10 bars.
29,30	Ball valves, quarter-turn, on-off operation, made of brass, PN 16 bars to BS 5154, threaded to BS 21, or ISO 7.
31	Filter (strainer), screen type, or grooved disks, 120 mesh/130 micron, epoxy coated metal body, or other high quality material, PN 10, complete with pressure inspection valves, wash-out drain valve, threaded connection to BS 21.
32	On-line, point-source dripper emitters, turbulence flow made of high quality plastic material, 24 litres/h discharge, 1.0 bar operating pressure, cv < 7%, filtration requirements 120 mesh/130 micron.
33	Pop-up sprinkler rotary gear driven full circle, 0.7–0.8 m^3/h discharge at 2–2.5 bars operating pressure, radius 7 m, interchangeable nozzle, c/with small strainer, drain mechanism and plastic cover, threaded (F) connection ¾ inch to BS 21.
35	Valve boxes, made of reinforced plastic or any other material, with cutout openings for pipe on opposite sides of the open bottom, c/with tight-fitting lids or covers on the top. Approximate dimensions: 33 cm *x* 45 cm (base) *x* 30 cm height.
36	The trench should be as uniform and level as possible, free of large stones and any other sharp-edged materials. Where required it must be filled with embedment material such as grained soil or sand to a depth of 10 cm. Trench dimensions should be 60 cm minimum depth for the 75 mm pipe and 50 cm for the 63 and 50 mm pipes, and 35 cm minimum width in all cases.

CHAPTER 6: Irrigation scheduling

Irrigation scheduling is one of the factors that influence the agronomic and economic viability of small farms. It is important for both water savings and improved crop yields. The irrigation water is applied to the cultivation according to predetermined schedules based upon the monitoring of:

- the soil water status;
- the crop water requirements.

The type of soil and climatic conditions have a significant effect on the main practical aspects of irrigation, which are the determination of how much water should be applied and when it should be applied to a given crop.

In addition to the basic factors relevant to the preparation of irrigation schedules examined below, other important elements should also be considered, such as crop tolerance and sensitivity to water deficit at various growth stages, and optimum water use.

SOIL-WATER RELATIONSHIP

Table 6.1 presents a summary table of soil physical properties.

TABLE 6.1 - Soil physical properties (average values)

Type of soil	Light (coarse) texture	Medium texture	Heavy (fine) texture
Saturation capacity (SC) % weight	25–35%	35–45%	55–65%
Field capacity (FC) % weight	8–10%	18–26%	32–42%
Wilting point (WP) % weight	4–5%	10–14%	20–24%
SC/FC	2/1	2/1	2/1
FC/WP	2/1	1.85/1	1.75/1
Bulk density(volume weight)	1.4–1.6 g/cm³	1.2–1.4 g/cm³	1.0–1.2 g/cm³
Soil available water (moisture) by volume (FC-WP *x* bulk density)	6%	12%	16–20%
Available moisture (Sa) in mm per metre soil depth (FC-WP *x* bulk density *x* 10)	60 mm	120 mm	160–200 mm
Soil water tension in bars:			
• at field capacity	0.1	0.2	0.3
• at wilting point	15.0	15.0	15.0
Time required from saturation to field capacity	18–24 h	24–36 h	36–89 h
Infiltration rate	25–75 mm/h	8–16 mm/h	2–6 mm/h

Example:
The field capacity (FC) of a 45-cm layer of soil is 18 percent. How much water in cubic metres per hectare does this layer hold?

Answer:
FC = 18 %, WP = FC ÷ 1.85 = 9.7 %, Sa = 18-9.7 = 8.3 %;
Bulk density = 1.2 g/cm^3; Sa mm/m = 8.3 x 1.2 x 10 = 99.6, Sa mm/45 cm = 8.3 x 1.2 x 10 x 0.45 = 44.8 mm; m^3/ha = 0.0996 ÷ 1 x 0.45 x 10 000 (1 ha) = 448.2, or m^3/ha = Sa (mm/m) x depth of layer (m) x 10.

Therefore, the answer is 448.2 m^3/ha.

EFFECTIVE ROOT DEPTH

This is the soil depth from which the plants take nearly 80 percent of their water needs, mostly from the upper part where the root system is denser. The rooting depths depend on the plant physiology, the type of soil, and the water availability (kind of irrigation). Indicative figures are presented in FAO Irrigation and Drainage Paper No. 24, Table 39.

In general, vegetables (beans, tomatoes, potatoes, onions, peanuts, cucumbers, etc.) are shallow rooted, about 50–60 cm; fruit trees, cotton and some other plants have medium root depths, 80–120 cm. Alfalfa, sorghum, and maize have deeper roots (Table 6.2). Moreover, rooting depths vary according to age.

TABLE 6.2 - Example of rooting depth (metres) during the growing season

	August	September	October	November	December	January
Maize	-	0.4	0.9	1.2	1.2	-
Cotton	0.4	0.8	1.0	1.0	1.0	-
Tomato	-	-	0.3	0.7	0.9	0.9

PERMISSIBLE DEFICIT OR DEPLETION OF SOIL AVAILABLE WATER

The fraction of moisture in the soil which amounts to 20–70 percent of the total available moisture (Sa) and is easily absorbed by the plants (without any stress that results in yield reduction) is called readily available moisture. It is a product of Sa multiplied by p, which represents the maximum permissible depletion of available water (moisture). The p value differs according to the kind of plant, the root depth, the climatic conditions and the irrigation techniques. Values for p are given in FAO Irrigation and Drainage Paper No. 33, Tables 19 and 20, and vary from

0.25 in shallow rooted sensitive crops to 0.70 in deep rooted tolerant crops. Table 23 of the same paper provides information on the sensitive growth periods of different crops.

Field observations have shown that the lower the soil moisture depletion (p), the better the crop development and yield. Hence, the recommended p values are:

- 0.20–0.30 for shallow rooted seasonal crops;
- 0.40–0.60 for deep rooted field crops and mature trees.

NET IRRIGATION APPLICATION DEPTH

Irrigation takes place when the permissible percentage (p) of available water (Sa) is depleted from the root depth, i.e. to replenish the depleted water. Therefore:

$$\text{Net depth of irrigation dose } (d)\ (mm) = (Sa \times p)\ D$$

where **Sa** is the available water in millimetres per metre, **p** is the permissible depletion (fraction), and **D** is the root depth (m).

Example:
Where Sa = 99 mm/m, p = 0.5, D = 0.4 m, what is the net irrigation dose (d) in millimetres to replenish the moisture deficit?

$$d = 99 \times 0.5 \times 0.4 = 19.8\ mm$$

CROP WATER REQUIREMENTS

The amount of water which evaporates from wet soils and plant surfaces together with the plant transpiration is called evapotranspiration (ET). Its value is largely determined by climate factors, such as solar radiation, temperature, humidity and wind, and by the environment. Out of the total evapotranspiration, evaporation accounts for about 10 percent and plant transpiration for the remaining 90 percent. Crop water requirements encompass the total amount of water used in evapotranspiration.

Alternative approaches for estimating the evapotranspiration, such as the radiation, Penman and pan methods, are presented in FAO Irrigation and Drainage Papers Nos. 24 and 33. Reference evapotranspiration (ETo) represents the rate of evapotranspiration of green grass under ideal conditions, 8–15 cm tall, with extensive vegetative cover completely shading the ground. It is expressed as a mean value in millimetres per day over a period of 10–30 d.

The most practical method for determining ETo is the pan evaporation method. This approach combines the effects of temperature, humidity, wind speed and sunshine. The best known pans are the Class A evaporation pan (circular) and the Colorado sunken pan (square).

The evaporation from the pan is very near to the evapotranspiration ofgrass that is taken as an index of ETo for calculation purposes. The pan direct readings (Epan) are related to the ETo with the aid of the pan coefficient (kpan), which depends on the type of pan, its location (surroundings with or without ground cover vegetation) and the climate (humidity and wind speed) (Table 6.3). Hence, ETo = Epan *x* kpan.

The kpan values for both types of pans are given in FAO Irrigation and Drainage Paper No. 24, Tables 18 and 19. For the Class A pan the average kpan is 0.70 and for the Colorado sunken pan it is 0.80.

Example:

TABLE 6.3 - Estimate of ETo in millimetres per day in the Wadi Tuban Delta

Month	June	July	August	September	October	November	December
Epan	9.0	8.8	8.8	8.2	8.0	6.5	5.7
kpan			average	0.70			
ETo	6.3	6.2	6.2	5.7	5.6	4.5	4.0

In order to relate ETo to crop water requirements (ETc), the specific crop coefficient (kc) must be determined: ETc = ETo *x* kc.

The crop coefficient (kc) depends on the crop leaf area and its roughness, the stage of growth, the growing season and the prevailing weather conditions (Table 6.4). Tables 6.5 and 6.6 list the kc values for different crops at various growth stages.

Example:

TABLE 6.4 - Cotton, growing season August-December

	August	September	October	November	December
ETo mm/d	6.2	5.7	5.6	4.5	4.0
Cotton kc	0.4	0.7	1.1	1.0	0.8
Cotton ET cmm/d	2.5	4.0	6.2	4.5	3.2
Cotton ET cmm/month	78	120	192	135	99

Total net water requirement approximately 580 mm
(December taken as half)

TABLE 6.5 - Crop factor (kc) for seasonal crops (average figures)

Crop	Initial	Crop development	Mid-season	Late and harvest
Bean (green)	0.35	0.70	1.0	0.9
Bean (dry)	0.35	0.75	1.1	0.5
Cabbage	0.45	0.75	1.05	0.9
Carrot	0.45	0.75	1.05	0.9
Cotton	0.45	0.75	1.15	0.75
Cucumber	0.45	0.70	0.90	0.75
Eggplant	0.45	0.75	1.15	0.80
Groundnut	0.45	0.75	1.0	0.75
Lettuce	0.45	0.60	1.0	0.90
Maize (sweet)	0.40	0.80	1.15	1.0
Maize (grain)	0.40	0.75	1.15	0.70
Melon	0.45	0.75	1.0	0.75
Onion (green)	0.50	0.70	1.0	1.0
Onion (dry)	0.50	0.75	1.05	0.85
Pea (fresh)	0.45	0.80	1.15	1.05
Pepper	0.35	0.75	1.05	0.90
Potato	0.45	0.75	1.15	0.75
Spinach	0.45	0.60	1.0	0.90
Squash	0.45	0.70	0.90	0.75
Sorghum	0.35	0.75	1.10	0.65
Sugar beet	0.45	0.80	1.15	0.80
Sugar cane	0.45	0.85	1.15	0.65
Sunflower	0.35	0.75	1.15	0.55
Tomato	0.45	0.75	1.15	0.80

TABLE 6.6 - Crop factor (kc) for permanent crops

Crop	Young	Mature
Banana	0.50	1.10
Citrus	0.30	0.65
Apple, cherry, walnut	0.45	0.85
Almond, apricot, pear, peach, pecan, plum	0.40	0.75
Grape, palm tree	0.70	0.70
Kiwi	0.90	0.90
Olive	0.55	0.55
Alfalfa	0.35	1.1

EFFECTIVE RAINFALL

In many areas, seasonal rain precipitation (P) might provide part of the water requirements during the irrigation season. The amount of rainwater retained in the root zone is called effective rainfall (Pe) and should be deducted from the total irrigation water requirements calculated. It can be roughly estimated as:

Pe = 0.8 P where P > 75 mm/month;
Pe = 0.6 P where P < 75 mm/month.

GROUND COVER

Another element to consider when estimating crop water requirements is the percentage of the field area (ground) covered by the cultivation. A reduction factor, expressed as kr, is applied to the conventional ET crop calculations. This factor is slightly higher, by about 15 percent, than the actual ground covered by the crop. For example, if the actual ground cover is 70 percent, kr = 0.70 x 1.15 = 0.80.

IRRIGATION INTERVAL OR FREQUENCY

This is the number of days between two consecutive irrigations, i = d ÷ ETc, where d is the net depth of irrigation application (dose) in millimeters and ETc is the daily crop evapotranspiration in millimetres per day.

Example:
Where d is 19.8 mm, and ETc is 2.5 mm/d, then i = 19.8 ÷ 2.5 = 8 days.

IRRIGATION APPLICATION EFFICIENCY

The amount of water to be stored in the root zone is estimated as the net irrigation dose (d). However, during the irrigation process, considerable water loss occurs through evaporation, seepage, deep percolation, etc. The amount lost depends on the efficiency of the system (Table 6.7). Irrigation field application efficiency is expressed as:

$$Ea = \frac{d}{Water\ Applied\ (gross)} \times 100$$

where, **d** is water stored in the rootzone and **Water Applied (gross)** is the irrigation water.

Example:
The net irrigation dose (d) for an area of 1 ha is 19.8 mm, i.e. 198 m^3. The water delivered during irrigation is 280 m^3. What is the application efficiency?

Answer:
Ea = 198 x 100 ÷ 280 = 70.7 percent, or expressed as a fraction, 0.70. The remaining 30 percent of water applied is lost.

TABLE 6.7 - Approximate application efficiency of various on-farm irrigation systems and methods

System/method	Ea %
Earth canal network surface methods	40–50
Lined canal network surface methods	50–60
Pressure piped network surface methods	65–75
Hose irrigation systems	70–80
Low-medium pressure sprinkler systems	75
Microsprinklers, micro-jets, minisprinklers	75–85
Drip irrigation	80–90

GROSS IRRIGATION APPLICATION DEPTH

Given the irrigation efficiency as a fraction, i.e. Ea = 0.60 (60 percent), the gross depth of irrigation application or gross irrigation dose (dg) is calculated as follows:

$$dg = \frac{d}{Ea\ (fraction)}$$

LEACHING REQUIREMENTS

The salinity level in the root zone is related directly to the water quality, the amount of fertilizers and the irrigation application depth. A high salt content in the soil is controlled by leaching (see Chapter 7 Water Quality). An excess amount of water, 10–15 percent, is applied during the irrigation where necessary for leaching purposes. In this way a portion of the water percolates through and below the root zone carrying with it a portion of the accumulated soluble salts. The leaching requirements (LR) are considered for the calculation of the gross irrigation application (d).

SYSTEM FLOW (SYSTEM CAPACITY)

The minimum flow capacity of any irrigation system should be the one that can meet the water requirements of the area under irrigation at peak demand:

$$minimum Q = 10\ A\ \frac{dg}{it}$$

where ***Q*** is the system flow in cubic metres per hour, ***A*** is the area in hectares, ***dg*** is the gross irrigation application depth (irrigation dose) in millimetres, ***i*** is the interval in days between two irrigations at peak demand, ***t*** is the operating hours per day, and 10 is a constant for hectares. However, the minimum flow of the system should be the one that enables the completion of irrigation at least two days before the next irrigation. This allows time to repair any damage to the system or pumping unit. Therefore, the value of i in the above formula should be reduced by two days.

The duration of application per irrigation is determined as:

$$T = 10\,A\,\frac{dg}{Q}$$

where ***T*** is the total operating hours of the system.

GENERAL EXAMPLE

In the following example (Table 6.8) the effective rainfall (Pe), the ground cover (kr) and the leaching requirements (LR) are not considered. However, these elements are important in localized micro-irrigation systems.

- Crop: Cotton
- Area: 1.5 ha.
- Location: Tuban Delta.
- Growing season: August-December.
- Irrigation method: Pressure piped surface method.
- Irrigation efficiency: 70 percent.
- Soil of medium texture, Sa = 99 mm/m.

TABLE 6.8 - Cotton example

	August	September	October	November	December
Soil available water Sa mm/m	99	99	99	99	99
Depletion of available water p	0.5	0.6	0.6	0.6	0.6
Cotton root depth D m	0.4	0.7	1.0	1.0	1.0
Net irrigation application d mm	19.8	41.6	59.4	59.4	59.4
Epan mm/d	8.8	8.2	8.0	6.5	5.7
kpan	0.7	0.7	0.7	0.7	0.7
ETo mm/d	6.2	5.7	5.6	4.5	4.0
Cotton kc	0.4	0.7	1.1	1.0	0.8
Cotton ETc mm/d	2.5	4.0	6.2	4.5	3.2
Irrigation interval i days	8	10.5	9.6	13	18.5
Gross irrigation dose dg mm	28.3	59.4	85.0	85.0	85.0
Gross irrigation dose dg m^3/h	425	891	1 275	1 275	1 275

The peak demand is in October when ETc is 6.2 mm/d and the irrigation frequency (interval) is 8 days. If the number of operating hours per day is seven, the system flow should be:

$$minimumQ = 10\ \frac{1.5ha \times 85mm}{(9days - 2days) \times 7hrs/day} = 26m^3/\mathrm{hr}$$

The duration of application per irrigation would be as follows (Table 6.9):

- August: T = 10 x 1.5 x 28.3 ÷ 26 = 16.3 hours, i.e. 2 days;
- September: T = 10 x 1.5 x 59.4 ÷ 26 = 34.3 hours, i.e. 5 days;
- October: T = 10 x 1.5 x 85.0 ÷ 26 = 49.0 hours, i.e. 7 days;
- November: T = 10 x 1.5 x 85.0 ÷ 26 = 49.0 hours, i.e. 7 days;
- December: T = 10 x 1.5 x 85.0 ÷ 26 = 49.0 hours, i.e. 7 days.

TABLE 6.9 - Irrigation programme

End of July	**pre-sowing irrigation to wet 0.6 m soil depth**	**1 273 m³**
Beginning of August	crop establishment	
8 August	irrigation	425 m³
16 August	irrigation	425 m³
24 August	irrigation	425 m³
1 September	irrigation	891 m³
11 September	irrigation	891 m³
22 September	irrigation	891 m³
2 October	irrigation	1 275 m³
11 October	irrigation	1 275 m³
21 October	irrigation	1 275 m³
31 October	irrigation	1 275 m³
13 November	irrigation	1 275 m³
26 November	irrigation	1 275 m³

The last irrigation on 26 November can last up to 9 December, i.e. until harvest. The total amount of water that must be applied to this crop on an area of 1.5 ha is: 11 598 m^3 plus 1 273 m^3 as the minimum amount for pre-irrigation, for a total of 12 871 m^3.

CHAPTER 7: Water quality for irrigation

INTRODUCTION

Irrigation waters whether derived from springs, diverted from streams, or pumped from wells, contain appreciable quantities of chemical substances in solution that may reduce crop yield and deteriorate soil fertility. In addition to the dissolve salts, which has been the major problem for centuries, irrigation water always carry substances derived from its natural environment or from the waste products of man's activities (domestic and industrial effluents). These substances may vary in a wide range, but mainly consist of dirt and suspended solids (SS) resulting into the emitters' blockages in micro-irrigation systems and bacteria populations and coliforms harmful to the humans and the animals.

CLASSIFICATION OF WATER QUALITY FOR IRRIGATION

In several studies carried out in the eighties on the various causes for the clogging of the emitters, the engineers, based on the three major interrelated factors contributing to that specific problem, classified the water according to the chemical quality, the physical and the biological. This classification although primary and simple seemed convenient for a more or less broad evaluation to cover the whole spectrum of irrigation waters quality for crop production. With the inclusion of reuse of treated municipal wastewater in agriculture during that period, the water quality considerations were broadened in order to cover all the physicochemical, biological and microbiological properties of water that may cause any impact on soil, plants, environment and the consumers, human or livestock. The water quality evaluation method in this chapter, although in brief, draws on the important parameters and criteria for a more or less practical evaluation of the chemical, the physical and the biological quality of the water for irrigation with pressurized techniques as follows:

a) Chemical, (salinity/toxicity hazards for the soil, the plants and the irrigation system such as it is pipe corrosion and emitter chemical clogging),
b) Physical (emitters blockages problems from suspended solid particles and other impurities content),
c) Biological (problems from bacteria and other contents harmful for human and animal health as well as for the soil the plants and the irrigation systems).

The information on the physical and biological evaluation is given in brief as compared to the chemical evaluation. Yet an in-depth and complete examination should include the soil physical properties and the climatic conditions as well as many other factors with direct or indirect influence on water use in agriculture and landscape.

THE CHEMICAL QUALITY OF IRRIGATION WATER

Composition and concentration of soluble salts

Salinity is a common problem facing farmers who irrigate in arid climates. This is because all irrigation waters contain soluble salts. Whether derived from springs, diverted from streams, or pumped from wells, the waters contain appreciable quantities of chemical substances in solution, dissolved from the geological strata through and over which the waters have flowed. Waters with a high salt content may have moved from a saline water table. In areas with intensive agriculture, fertilization is a major cause of aquifer salinization.

The composition of salts in water varies according to the source and properties of the constituent chemical compounds. These salts include substances such as gypsum (calcium sulphate, $CaSO_4.2H_2O$), table salt (sodium chloride NaCl) and baking soda (sodium bicarbonate $NaHCO_3$). When dissolved in water, salts separate into ions; e.g. sodium chloride breaks down into sodium and chloride ions. Thus, it is customary to refer to ions rather than salts. The principle ions in irrigation water and their characteristics are listed in Table 7.1.

TABLE 7.1 - Principle ions present in irrigation water

Ions	Chemical symbol	Equivalent weight
Anions (acidic ions)		
Chloride	Cl^-	35.5
Sulphate	SO_4^{--}	48
Carbonate	CO_3^{--}	30
Bicarbonate	HCO_3^-	61
Nitrate	NO_3^-	62
Cations (basic ions)		
Sodium	Na+	23
Potassium	K+	39.1
Calcium	Ca++	20
Magnesium	Mg++	12.2

All ions are expressed in the form of milligrams per litre (mg/litre or ppm) and milliequivalents per litre (meq/litre). The latter unit is preferable because water quality criteria involve milliequivalents per litre calculations.

The conversion formula is:

$$meq/litre = \frac{mg/litre}{equivalentweight}$$

Boron is also present in irrigation waters as un-ionized boric acid expressed as boron element (B) in milligrams per litre. The salt concentration in most irrigation waters ranges from 200 to 4 000 mg/litre total dissolved solids (TDS). The pH of the water is also an indicator of its quality and it normally ranges from 6.5 to 8.4.

The common method for evaluating the total salts content in water is by measuring the electrical conductivity of water (ECw) at 25°C. Electrical conductivity is expressed in deciSiemens per metre. There is a relation between the electrical conductivity and the concentration of salts in milliequivalents per litre and in milligrams per litre when the ECw is in the range of 1–5 dS/m. Thus, every 10 meq/litre of salts (cation concentration) create 1 dS/m ECw. The relationship between electrical conductivity and total dissolved salts (TDS) is:

$$ECw\ (dS/m) \times 640 = TDS\ (mg/litre)$$

The sum of cations should equal the sum of anions. The accuracy of the chemical water analyses should be checked on the basis of the above relationships.

Effect of soluble salts on plants

The application of irrigation water to the soil introduces salts into the root zone. Plant roots take in water but absorb very little salt from the soil solution. Similarly, water evaporates from the soil surface but salts remain behind. Both processes result in the gradual accumulation of salts in the root zone, even with low salinity water. This situation may affect the plants in two ways: a) by creating salinity hazards and water deficiency; and b) by causing toxicity and other problems.

Salinity hazards and water deficiency

The build-up of salinity in the root zone increases the osmotic pressure of the soil solution and causes a reduction in both the rate of water absorption by the plants and the soil water availability. Thus, a continuous water deficiency may exist even though the field is heavily irrigated. Plant wilting symptoms may not become apparent, but growth and yield are depressed. Under such circumstances it is not possible to maintain good crop development conditions and obtain high yields. Instead, plant growth is delayed and there is a considerable reduction in yield. Seed germination is also affected by the presence of salts. It is usually delayed and in some cases does not occur.

The level of salinity build-up depends on both the concentration and the composition of salts in the water. Chloride is highly soluble and remains in the soil solution, while sulphate and bicarbonate combine with calcium and magnesium, where present, to form calcium sulphate and calcium carbonate, which are sparingly soluble compounds.

Toxicity hazards

Many fruit trees and other cultivations are susceptible to injury from salt toxicity. Chloride, sodium and boron are absorbed by the roots and transported to the leaves where they accumulate. In harmful amounts, they result in leaf burn and leaf necrosis. Moreover, direct contact during sprinkling of water drops with a high chloride content may cause leaf burn in high evaporation conditions. To some extent, bicarbonate is also toxic. Other symptoms of toxicity include premature leaf drop, reduced growth and reduced yield. In most cases, plants do not show clear toxicity problems until it is too late to remedy the situation.

Chloride and sodium ions are both present in the solution. Thus, it is difficult to determine whether the damage caused is due to the one or to the other. Chloride ions in high concentrations are known to be harmful to citrus and many woody and leafy field crops. A chloride content exceeding 10 meq/litre may cause severe problems to crops. The effect of sodium toxicity is not very clear. However, it has been found that it may cause some direct or indirect damage to many plants.

Boron is an essential element to the plants. However, where present in excessive amounts, it is extremely toxic, even at relatively very low concentrations of 0.6 mg/litre. Toxicity occurs with the uptake of boron from the soil solution. The boron tends to accumulate in the leaves until it becomes toxic to the leaf tissue and results in the death of the plant. In arid regions, boron is considered the most harmful element in irrigation water.

Other problems

In addition to the moisture availability effect and the toxicity problems to which the soluble salts contribute, certain salt constituents may interfere with the normal nutrition of various crops. Bicarbonate ions in high concentrations may affect the uptake of mineral nutrients and their metabolism in the plant. Chlorotic symptoms in sensitive plants may be due to the direct or indirect effects of bicarbonate, e.g. an increase in soil pH.

Excessive nitrate contents, higher than 100 mg/litre, may affect transplants and sensitive crops at the initial growth stage. However, no negative effects have been reported in the last three decades from fertigation with pure nitrogen concentrations in irrigation water of about 200 ppm. Although there is no doubt about the problem's existence, it seems that the main concern should be the nitrate content in the irrigation water, when calculating the total nitrogen application, NO_3 equals 0.226 N (pure nitrogen).

Effects of soluble salts on soil

Sodium hazard

A soil permeability problem occurs with a high sodium content in the irrigation water. Sodium has a larger concentration than any other cation in saline water, its salts being very soluble. Positively charged, it is attracted by negatively charged soil particles, replacing the dominant calcium and magnesium cations. The replacement of the calcium ions with sodium ions causes the dispersion of the soil aggregates and the deterioration of its structure, thus rendering the soil impermeable to water and air. The increase in the concentration of exchangeable sodium may cause an increase in the soil pH to above 8.5 and reduce the availability of some micronutrients, e.g. iron and phosphorus.

The degree of absorption to the clay particles of the sodium depends on its concentration in the water and the concentration of the calcium and magnesium ions. This reaction is called cation exchange and it is a reversible process. The capacity of soil to adsorb and exchange cations is limited. The percentage of the capacity that sodium takes up is known as the exchangeable sodium percentage (ESP). Soils with ESP > 15 are seriously affected by adsorbed sodium.

The sodium problem is reduced if the amount of calcium plus magnesium is high compared with the amount of sodium. This relation is called the sodium adsorption ratio (SAR) and it is a calculated value from the formula:

$$SAR = \frac{Na}{\sqrt{\frac{Ca + Mg}{?}}}$$

(ions units meq/litre)

The use of water with a high SAR value and low to moderate salinity may be hazardous and reduce the soil infiltration rate. The SAR of irrigation water indicates the approximate ESP of a soil with water.

Residual sodium carbonate (RSC)

This is defined as the difference in milliequivalents per litre between the bicarbonate ions and those of calcium and magnesium. Calcium and magnesium may react with bicarbonate and precipitate as carbonates. The relative sodium concentration in the exchangeable complex increases resulting in the dispersion of soil. When the RSC value is lower than 1.25 meq/litre, the water is considered good quality, while if the RSC value exceeds 2.5 meq/litre, the water is considered harmful.

Crop tolerance to salinity

Crop tolerance is the degree to which a crop can grow and yield satisfactorily in saline soil. Different crops vary widely in their response to salinity. Some can tolerate less than 2 dS/m and others up to and above 8 dS/m. Salt tolerance also depends considerably upon cultural conditions and irrigation management practices. Many other factors such as plant, soil, water and climate interact to influence the salt tolerance of a crop.

Relative salt tolerance data have been developed for many crops and are used as general guidelines. The following data are related to the expected decline in yield. The ECe is the soil salinity in terms of electrical conductivity (EC) measured from the soil saturation extract, with a value of 1.5 EC for irrigation water (ECiw). Tables 13–18 (taken from Maas, 1990) give two important parameters for expressing a plant's salt tolerance:

- Threshold - the maximum allowable salinity of soil saturation extract (ECe).
- Slope - the percent yield decrease per unit increase in salinity.

The rating of plants according to their sensitivity/tolerance to salts (Tables 7.2, 7.3, 7.4, 7.5, 7.6 and 7.7) is even more important as it provides vital information at first sight for the evaluation of and diagnosis for potential salinity problems.

TABLE 7.2 - Relative salt tolerance of herbaceous crops - vegetables and fruit crops (Maas, 1990)

Common name	Botanical name	Threshold dS/m	Slope % per dS/m	Rating
Artichoke	*cynara scolymus*	--	--	MT*
Asparagus	*asparagus officinalis*	4.1	2.0	T
Bean	*phaseolus vulgaris*	1.0	19.0	S
Bean, mung	*vigna radiata*	1.8	20.7	S
Beet, red	*beta vulgaris*	4.0	9.0	MT
Broccoli	*brassica oleracea botrytis*	2.8	9.2	MS
Brussels sprouts	*b. oleracea gemmifera*	--	--	MS*
Cabbage	*b. oleracea capitata*	1.8	9.7	MS
Carrot	*daucus carota*	1.0	14.0	S
Cauliflower	*brassica oleraca botrytis*	--	--	MS*
Celery	*apium graveolens*	1.8	6.2	MS
Corn, sweet	*zea mays*	1.7	12.0	MS
Cucumber	*cucumis sativa*	2.5	13.0	MS
Eggplant	*solanum melongena esculentum*	1.1	6.9	MS
Kale	*brassica oleracea acephala*	--	--	MS*
Kohlrabi	*b. oleracea gongylode*	--	--	MS*
Lettuce	*lactuca sativa*	1.3	13.0	MS
Muskmelon	*cucumis melo*	--	--	MS
Okra	*abelmoschus esculentus*	--	--	S
Onion	*akkium cepa*	1.2	16.0	S
Parsnip	*pastinaca sativa*	--	--	S*
Pea	*pisum sativa*	--	--	S*
Pepper	*capsicum annuum*	1.5	14.0	MS
Potato	*solanum tuberosum*	1.7	12.0	MS
Pumpkin	*cucurbita pepo pepo*	--	--	MS*
Radish	*raphanus sativus*	1.2	13.0	MS
Spinach	*spinacia oleracea*	2.0	7.6	MS
Squash scallop	*curcubita melo melopepo*	3.2	16.0	MS
Squash zucchini	*curcubita melo melopepo*	4.7	9.4	MT
Strawberry	*fragaria sp.*	1.0	33.0	S
Sweet potato	*ipomoea batatas*	1.5	11.0	MS
Tomato	*lycopersicon lycopersicum*	2.5	9.9	MS
Tomato cherry	*l.esculentum var cerasiforme*	1.7	9.1	MS
Turnip	*brassica rapa*	0.9	9.0	MS
Watermelon	*citrullus lanatus*	-	--	MS*

Note:

S sensitive, MS moderately sensitive, MT moderately tolerant, T tolerant. The above data serve only as a guideline to relative tolerance among crops. Absolute tolerance varies, depending upon climate, soil conditions, and cultural practices.

In gypsipherus soils, plants will tolerate an ECe about 2 dS/m higher than indicated.

***:** *Ratings are estimates.*

TABLE 7.3 - Relative salt tolerance of herbaceous crops - woody crops (Maas, 1990)

Common name	Botanical name	Threshold dS/m	Slope % per dS/m	Rating
Almond**	*prunus duclis*	1.5	19.0	S
Apple	*malus sylvestris*	--	--	S
Apricot**	*prunus americana*	1.6	24.0	S
Avocado**	*persea americana*	--	--	S
Blackberry	*rubus sp*	1.5	22.0	S
Boysenberry	*rubus ursinus*	1.2	22.0	S
Castor seed	*ricinus communus*	--	--	MS*
Cherimoya	*annona cherimola*	--	--	S*
Cherry, sweet	*prunus avium*	--	--	S*
Cherry, sand	*prunus besseyi*	--	--	S*
Currant	*ribes sp.*	--	--	S*
Date palm	*phoenix dactylifera*	4.0	3.6	T
Fig	*ficus carica*	--	--	MT*
Gooseberry	*ribes sp.*	--	--	S*
Grape**	*vitis sp.*	1.5	9.6	MS
Grapefruit**	*citrus paradisi*	1.8	16.0	S
Guayule	*parthenium argentantum*	15.0	13.0	T
Jojoba**	*simmondsia chinensis*	--	--	T
Jujube	*ziziphus jujuba*	--	--	MT*
Lemon**	*citrus limon*	--	--	S
Lime	*citrus aurantiifolia*	--	--	S*
Loquat	*eriobotrya japonica*	--	--	S*
Mango	*mangifera indica*	--	--	S*
Olive	*olea europea*	--	--	MT
Orange	*citrus sinensis*	1.7	16.0	S
Papaya**	*carica papaya*	--	--	MT
Passion fruit	*passiflora edulis*	--	--	S*
Peach	*prunus persica*	1.7	21.0	S
Pear	*pyrus communis*	--	--	S*
Persimmon	*diospyros virginiana*	--	--	S*
Pineapple	*ananas comosus*	--	--	MT*
Plum, prune**	*prunus domestica*	1.5	18.0	S
Pomegranate	*punica granatum*	--	--	MT*
Pummelo	*citrus maxima*	--	--	S*
Raspberry	*rubus idaeus*	--	--	S
Rose apple	*syzygium jambos*	--	--	S*
Sapote, white	*casimiroa edulis*	--	--	S*
Tangerine	*citrus reticulata*	--	--	S*

Note:

Data applicable when rootstocks are used that do not accumulate Na or Cl rapidly, or when these ions do not predominate in the soil.

In gypsipherus soils, plants will tolerate an ECe about 2 dS/m higher than indicated.

***:** *Ratings are estimates.*
****:** *Tolerance is based on growth rather than yield*

TABLE 7.4 - Relative salt tolerance of herbaceous crops - grasses and forage crops (Maas, 1990)

Common name	Botanical name	Threshold dS/m	Slope % per dS/m	Rating
Alfalfa	*medicago sativa*	2.0	7.3	MS
Alkali sacaton	*sporobolus airoides*	--	--	T*
Alkaligrass, Nuttal	*puccinellia airoides*	--	--	T*
Barley (forage)	*hordeum vulgare*	6.0	7.1	MT
Barnet	*poterium sanguisorba*	--	--	MS*
Bentgrass	*agrostis stolonifera palustris*	--	--	MS
Bermudagrass	*cynodon dactylon*	6.9	6.4	T
Bluestem, Angleton	*dichanthium aristatum*	--	--	MS*
Brome, mountain	*bromus marginatus*	--	--	MT*
Brome, smooth	*b. inermis*	--	--	MS
Buffelgrass	*cenchrus ciliaris*	--	--	MS*
Canarygrass, reed	*phalaris arundinacea*	--	--	MT
Clover, alsike	*trifolium hybridum*	1.5	12.0	MS
Clover, Berseem	*trifolium alexandrinum*	1.5	5.7	MS
Clover, Hubam	*melitotus alba*	--	--	MT*
Clover, ladino	*trifolium repens*	1.5	12.0	MS
Clover, red	*trifolium pratense*	1.5	12.0	MS
Clover, strawberry	*melitotus*	--	--	MT*
Clover, white Dutch	*trifolium repens*	--	--	MS*
Corn (forage)	*zea mays*	1.8	7.4	MS
Cowpea (forage)	*vigna unguiculata*	2.5	11.0	MS
Dallisgrass	*paspalum dilatatum*	--	--	MS*
Fescue, meadow	*f. pratensis*	--	--	MT*
Fescue, tall	*festuca elatior*	3.9	5.3	MT
Foxtail, meadow	*alopecurus pratensis*	1.5	9.6	MS
Grama, blue	*bouteloua gracilis*	--	--	MS*
Hardinggrass	*phalaris tuberosa*	4.6	7.6	MT
Kallargrass	*diplachne fusca*	--	--	T*
Lovergrass	*eragrostis sp.*	2.0	8.4	MS
Milkvetch, Cicer	*astragalus cicer*	--	--	MS*
Oatgrass, tall	*arrhenatherum danthonia*	--	--	MS*
Oats (forage)	*avena sativa*	--	--	MS*
Orchardgrass	*dactylis glomerata*	1.5	6.2	MS
Panicgrass, blue	*panicum antidotale*	--	--	MT*
Rape	*brassica napus*	--	--	MT*
Rescuegrass	*bromus unioloides*	--	--	MT*
Rhodesgrass	*chloris gayana*	--	--	MT
Rye (forage)	*secale sereale*	--	--	MS*
Ryegrass, Italian	*lolium italicum multiflorum*	--	--	MT*
Ryegrass, perennial	*l. perenne*	5.6	7.6	MT
Saltgrass, desert	*distichlis stricta*	--	--	T*
Sesbania	*sesbania exaltata*	2.3	7.0	MS
Sirato	*macroptilium atropurpureum*	--	--	MS
Sphaerophysa	*sphaerophysa salsula*	2.2	7.0	MS
Sudangrass	*sorghum sudanense*	2.8	4.3	MT
Timothy	*phleum pratense*	--	--	MS*
Trefoil, big	*lotus uliginosus*	2.3	19.0	MS
Trefoil, broadleaf	*l. corniculatus arvenis*	--	--	MT
Trefoil, narrowleaf	*l. corniculatus tenuifolium*	5.0	10.0	MT
Vetch, common	*vicia angustifolia*	3.0	11.0	MS
Wheat (forage)	*triticum aestivum*	4.5	2.6	MT
Wheat, durum (forage)	*t. turgidum*	2.1	2.5	MT

TABLE 7.4 - Relative salt tolerance of herbaceous crops - grasses and forage crops (cont'd)

Common name	Botanical name	Threshold dS/m	Slope % per dS/m	Rating
Wheatgrass, fairway	*a. cristatum*	7.5	6.9	T
Wheatgrass, intermediate	*a. intermedium*	--	--	MT*
Wheatgrass, slender	*a. trachycaulum*	--	--	MT
Wheatgrass, standard	*agropyron sibiricum*	3.5	4.0	MT
Wheatgrass, tall	*a. elongatum*	7.5	4.2	T
Wheatgrass, western	*a.smithii*	--	--	MT*
Wildrye, Altai	*elymus angustus*	--	--	T
Wildrye, beardless	*e. triticoides*	2.7	6.0	MT
Wildrye, Canadian	*e. canadensis*	--	--	MT*
Wildrye, Russian	*e. junceus*	--	--	T

Note:

The above data serve only as a guideline to relative tolerance among crops. Absolute tolerance varies, depending upon climate, soil conditions, and cultural practices.

In gypsipherus soils, plants will tolerate an ECe about 2 dS/m higher than indicated.

TABLE 7.5 - Boron tolerance limits for agricultural crops (Maas, 1990)

Common name	Botanical name	Threshold ** mg/litre	Slope % per mg/litre
Very sensitive			
Lemon*	*Citrus limon*	--	--
Blackberry	*Rubus sp.*	--	--
Sensitive			
Avocado	*persea american*	0.5–0.75	--
Grapefruit*	*citrus paradisi*	0.5–0.75	--
Orange*	*c. sinensis*	0.5–0.75	--
Apricot*	*prunus americana*	0.5–0.75	--
Peach*	*p. persica*	0.5–0.75	--
Cherry*	*p. avium*	0.5–0.75	--
Plum*	*p. domestica*	0.5–0.75	--
Persimmon*	*diospyros kaki*	0.5–0.75	--
Fig, kadota*	*ficus carica*	0.5–0.75	--
Grape*	*vitis vinifera*	0.5–0.75	--
Walnut*	*juglans regia*	0.5–0.75	--
Pecan*	*carya illinoienis*	0.5–0.75	--
Onion	*allium cepa*	0.5–0.75	--
Garlic	*allium sativum*	0.75–1.0	--
Sweet potato	*ipomea batatas*	0.75–1.0	--
Wheat	*triticum aestivum*	0.75–1.0	0.33
Sunflower	*helianthus annuus*	0.75–1.0	--
Bean, mung*	*vigna radiata*	0.75–1.0	--
Sesame*	*sesamum indicum*	0.75–1.0	--
Lupine*	*lipinus hartwegii*	0.75–1.0	--
Strawberry*	*fragaria ap.*	0.75–1.0	--
Artichoke, Jerusalem*	*helianthus tuberosus*	0.75–1.0	--
Bean, kidney*	*phaseolus vulgaris*	0.75–1.0	--
Bean, limab	*p. lunatus*	0.75–1.0	--
Peanut	*arachis hypogaea*	0.75–1.0	--

***:** *Ratings are estimates.*

TABLE 7.5 - Boron tolerance limits for agricultural crops (cont'd)

Common name	Botanical name	Threshold ** mg/litre	Slope % per mg/litre
Moderately sensitive			
Broccoli	*brassica oleracea botrytis*	1.0	1.8
Pepper, red	*capsicum annuum*	1.0–2.0	--
Pea*	*pisum sativa*	1.0–2.0	--
Carrot	*daucus carota*	1.0–2.0	--
Radish	*raphsnus sativus*	1.0–2.0	1.4
Potato	*solanum tuberosum*	1.0–2.0	--
Cucumber	*cucumis sativus*	1.0–2.0	--
Moderately tolerant			
Cabbage*	*brassica oleracea capitata*	2.0–4.0	--
Turnip	*b. rapa*	2.0–4.0	--
Bluegrass, Kentucky*	*poa pratensis*	2.0–4.0	--
Barley	*hordeum vulgare*	3.4	4.4
Cowpea	*vigna unguigulata*	2.5	12
Oats	*avena sativa*	2.0–4.0	--
Corn	*zea mays*	2.0–4.0	--
Artichoke*	*cynara scolymus*	2.0–4.0	--
Tobacco*	*nicotiana tabacum*	2.0–4.0	--
Mustard*	*brassica juncea*	2.0–4.0	--
Clover, sweet*	*melitotus indica*	2.0–4.0	--
Squash	*cucurbita pepo*	2.0–4.0	--
Muskmelon*	*cucmis melo*	2.0–4.0	--
Cauliflower	*b. oleracea botrytis*	2.0–4.0	1.9
Tolerant			
Alfalfa*	*medicago sativa*	4.6–6.0	--
Vetch, purple*	*vicia bengalensis*	4.6–6.0	--
Parsley*	*petroselinum crispum*	4.6–6.0	--
Beet, red	*beta vulgaris*	4.6–6.0	--
Sugar beet	*b. vulgaris*	4.9	4.1
Tomato	*lycopersicum*	5.7	3.4
Very tolerant			
Sorghum	*sorhgum bicolor*	7.4	4.7
Cotton	*gossypium hirsutum*	6.0–10.0	--
Celery*	*apium graveolens*	9.8	3.2
Asparagus*	*asparagus officinalis*	10.0–15.0	--

***:** *Tolerance based on reduction in vegetative growth*
****:** *Maximum permissible concentration in soil water without reduction in yield. Boron tolerances may vary, depending upon climate, soil conditions, and crop varieties.*

TABLE 7.6 - Salt tolerance of ornamental shrubs, trees and ground cover (Maas, 1990)

Common name	Botanical name	Max. permissible ECe dS/m
Very sensitive		
Star jasmine	*Trachelospermum jasminoides*	1–2
Pyrenees cotoneaster	*Cotoneaster congestus*	1–2
Oregon grape	*Mahonia aquifolium*	1–2
Photinia	*Photinia fraseri*	1–2
Sensitive		
Pineapple guava	*feijoa sellowiana*	2–3
Chinese holly, cv. Burford	*Ilex cornuta*	2–3
Rose cv. Grenoble	*Rosa sp.*	2–3
Glossy abelia	*Abelia grandiflora*	2–3
Southern yew	*Podocarpus macrophyllus*	2–3
Tulip tree	*Liriodendron tulipifera*	2–3
Algerian ivy	*Hedera canariensis*	3–4
Japanese pittosporum	*Pittosporum tobira*	3–4
Heavenly bamboo	*Nandina domestica*	3–4
Chinese hibiscus	*Hibiscus rosa sinensis*	3–4
Laurustinus cv. Robustum	*Viburnum tinusm*	3–4
Strawberry tree, cv compact	*Arbutus unedo*	3–4
Grape myrtle	*Lagerstroemia indica*	3–4
Moderately sensitive		
Glossy privet	*Ligustrum lugidum*	4–6
Yellow sage	*Lantana camara*	4–6
Orchid tree	*Bauhinia purpurea*	4–6
Southern magnolia	*Magnolia grandiflora*	4–6
Japanese boxwood	*Buxus microphylla var. japonica*	4–6
Xylosma	*Xylosma congestum*	4–6
Japanese black pine	*Pinus thunbergiana*	4–6
Indian hawthorn	*Raphiolepis indica*	4–6
Dodonaea, cv. atropurpurea	*Dodonaea viscosa*	4–6
Oriental arborvitae	*Platycladus orientalis*	4–6
Thorny elaeagnus	*Elaeagnus pungens*	4–6
Spreading juniper	*Uniperus chinensis*	4–6
Pyracantha, cv. Graberi	*Pyracantha fortuneana*	4–6
Cherry plum	*Prunus cerasifera*	4–6
Moderately tolerant		
Weeping bottlebrush	*Callistemon viminalis*	6–8
Oleander	*Nerium oleander*	6–8
European fan palm	*Chamerops humilis*	6–8
Blue dracaena	*Cordiline indivisa*	6–8
Rosemary	*Rosmarinus officinalis*	6–8
Aleppo pine	*Pinus halepensis*	6–8
Sweet gum	*Liquidamabar styraciflua*	6–8
Tolerant		
Brush cherry	*Syzygium paniculatum*	> 8
Ceniza	*Leucophyllum frutescens*	> 8
Natal plum	*Carssa grandiflora*	> 8
Evergreen pear	*Pyrus Kawakamii*	> 8
Bougainvillaea	*Bougainvillaea spectabilis*	> 8
Italian stone pine	*Pinus pinea*	> 8
Very tolerant		> 10
White iceplant	*Delosperma alba*	> 10
Rosea iceplant	*Drosanthemum hispidum*	> 10
Purple iceplant	*Labranthus productus*	> 10
Croceum iceplant	*Hymenocyclus croceus*	> 10

Note:
Salinities exceeding the maximum permissible ECe may cause leaf burn, loss of leaves and/or excessive stunting.

TABLE 7.7 - Boron tolerance limits for ornamentals (Maas, 1990)

Common name	Botanical name	Threshold mg/litre
Very sensitive		
Oregon grape	*Mahonia aquifolium*	
Photinia	*Photinia x fraseri*	
Xylosma	*Xylosma congestum*	
Thorny elaeagnus	*Elaeagnus pungens*	
Laurustinus	*Viburnum tinus*	
Wax-leaf privet	*Ligustrum japonicum*	
Pineapple guava	*Feijoa sellowiana*	
Spindle tree	*Euonymu japonica*	
Japanese pittosporum	*Pittosporum tobira*	
Chinese holly	*Ilex cornuta*	
Juniper	*Juniperus chinensis*	
Yellow sage	*Lantana camara*	
American elm	*Ulmus americana*	
Sensitive		
Zinnia	*Zinnia elaeagnus*	0.5–1.0
Pansy	*Viola tricolor*	0.5–1.0
Violet	*Viola odorata*	0.5–1.0
Larkspur	*Delphinum sp.*	0.5–1.0
Glossy abelia	*Abelia x grandiflora*	0.5–1.0
Rosemary	*Rosmarinus officinalis*	0.5–1.0
Oriental arborvitae	*Platycladus orientalis*	0.5–1.0
Geranium	*Pelargoium x hortorum*	0.5–1.0
Moderately sensitive		
Gladiolus	*Gladiolus sp.*	1.0–2.0
Marigold	*Calendula officinalis*	1.0–2.0
Poinsettia	*Euphorbia pulcherrima*	1.0–2.0
China aster	*Callistephus chinensis*	1.0–2.0
Gardenia	*Gardenia sp.*	1.0–2.0
Southern yew	*Podocarpus macrophyllus*	1.0–2.0
Brush cherry	*Syzygium paniculatum*	1.0–2.0
Blue dracaena	*Cordyline indivisa*	1.0–2.0
Ceniza	*Leucophyllus frutenscens*	1.0–2.0
Moderately tolerant		
Bottlebrush	*Callistemon citrinus*	2.0–4.0
California poppy	*Eschscholzia californica*	2.0–4.0
Japanese boxwood	*Buxus microphylla*	2.0–4.0
Oleander	*Nerium oleander*	2.0–4.0
Chinese hibiscus	*Hibiscus rosa-senensis*	2.0–4.0
Sweet pea	*Lathyrus odoratus*	2.0–4.0
Carnation	*Dianthus caryophyllus*	2.0–4.0
Tolerant		
Indian hawthorn	*Raphiolepis indica*	6.0–8.0
Natal plum	*Carissa grandiflora*	6.0–8.0
Oxalis	*Oxalis bowiei*	6.0–8.0

Note:

Species listed in order of increasing tolerance based on appearance as well as growth reductions. Boron concentration exceeding threshold may cause leaf burn and leaf loss.

Water quality criteria

There have been calls to establish standards as a guide for judging the suitability of water for irrigation. Any classification should be based on the total concentration and the composition of salts. However, the suitability of water for irrigation also depends on other associated factors, such as the crop, soil, climate and management practices. The classification adopted by FAO in 1985 (after Maas), and proposed as an initial guide (Table 7.8), has proved most practical and useful in assessing water quality for on-farm water use. The principal parameters for water classification (crop response to salinity, sodium hazard and toxicity) are quite clear and understood by both the extension engineers and the farmers themselves for proper irrigation management and follow-up purposes.

With the FAO assessment method, the parameters taken into consideration are the four presented below.

Total salinity

TABLE 7.8 - Water classification by salinity

	EC dS/m	TDS mg/litre
Non-saline water	< 0.7	< 500
Saline water	0.7–42	500–30 000
Slightly saline	0.7–3.0	500–2 000
Medium saline	3.0–6.0	2 000–4 000
Highly saline	> 6.0	> 4 000
Very saline	> 14.0	> 9 000
Brine	> 42	> 30 000

Crop response to salinity

The Figure 7.1 below shows the expected yield reduction for each crop in accordance with its sensibility/tolerance to salt. This graph enables a quick assessment of the two main parameters for the water suitability.

FIGURE 7.1 - Divisions for relative salt tolerance ratings of agricultural crops (Maas, 1984)

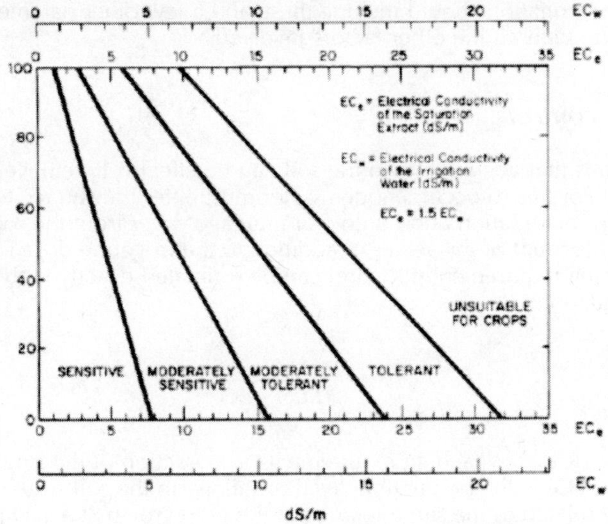

Sodium hazard

The sodium adsorption ratio is commonly used as an index of the sodium hazard of soils and waters, and as a substitute soil ESP. The SAR (Sodium Absorption Ratio) of a given water determines, to a certain extent, the relative amount of sodium that may be adsorbed by the soil. The effect of sodium ions in the irrigation water in reducing the infiltration rate and soil permeability is dependent on the total salt concentration, as shown in Table 7.9.

TABLE 7.9 - Potential infiltration problem due to sodium in irrigation water

Salinity levels of irrigation water dS/m	No reduction	Slight reduction	Medium reduction	Severe reduction
	SAR	SAR	SAR	SAR
ECw = 0.7	< 1	1–5	5–11	> 11
ECw = 0.7–3.0	< 10	10–15	15–23	> 23
ECw = 3.0–6.0	< 25	> 25	No effect	No effect
ECw = 6.0–14.0	< 35	> 35	No effect	No effect
ECw = >14.0	No effect	No effect	No effect	No effect

Source: Based on Rhoades, Oster and Schroer.

Toxicity problems

Toxicity problems may be created by excess chloride, sodium, boron, bicarbonate, nitrates and an abnormal pH. The evaluation of the water quality for irrigation should include these and a few other parameters in association with all the other factors involved.

Salinity control

The salts that accumulate in the soil can be effectively removed only by leaching. For this to occur, enough water must enter the surface to produce downward percolation and outflow of drainage water from the root zone. The extra amount of this water in addition to the irrigation dose is called the leaching requirement (LR), and can be estimated exactly with the use of the equation:

$$LR = \frac{ECw}{5(ECe) - ECw}$$

where **LR** is the leaching requirements as a fraction of the irrigation dose, and **ECe** is the permissible level of salinity in the soil solution primarily related to the salt tolerance of the crop grown at a 100 percent yield potential. The average value usually taken for ECe is 1.5 ECw. In this case, LR = 0.15.

Leaching is especially necessary as a soil preparation for crops with high plant density, such as carrots, onions and groundnuts. The salinity over the entire area should be the same with no difference between the wetted and the non-wetted parts of the field during the preceding season. The leaching of the salts in the top layer is particularly important because crops are sensitive to salinity during the first stages of their growth.

For the control of the salinity level in the root zone, frequent observations should be conducted with soil sampling for the laboratory determination of the soil extract EC. The use of soil solutions, extractors and portable metering devices on the spot enables the continuous monitoring, for immediate action, of any significant change in the EC of the soil solution, the chloride and nitrate content, and the soil pH as a result of irrigation and fertilization.

Micro-irrigation and salinity control

In drip irrigation, the distribution of dissolved salts in the soil profile follows the pattern of the water flux with the tendency for accumulation at the periphery of the wetted soil mass. Most of the wetted zone below the emitter, where most of the roots concentrate and function, remains free from salts during the irrigation season with low to medium salinity values. Near the surface, due to evaporation, the salt accumulation is five times greater than in the deeper layers and increases with distance from the emitters. This, in combination with the use of poor quality irrigation water and the application of fertilizers through the system, will cause a salinity build-up, which might become a problem in areas where the annual rainfall does not exceed 250 mm. In these cases, it is essential to flood the total area once a year, at the end of the season, with adequate amounts of water in order to leach the salts beyond the rooting depth.

The salinity level in the root zone is related to the water quality, the amount of fertilizers and the irrigation dose. The salt accumulation in the vicinity of the emitters is less than half that between the emitter lines. The EC value of the saturation extract beyond the emitter is 2–3 times the ECw, and between the lines it is six to ten times higher. This high salt content can be controlled only by leaching or by reducing the amount of fertilizer during the growing season. In no case should the fertilizer concentration in the irrigation water exceed EC 0.5 dS/m that is added to the total salinity of the irrigation water.

In drip irrigation, extra leaching with increased quantities of water every application during the irrigation season is not recommended unless salt accumulation reaches hazardous levels. Leaching should take place after the crop harvest, between irrigation seasons, where the salt content is excessive and the rainfall is not sufficient. It is done either by flooding the area or by low precipitation sprinklers with very fine drops (Tables 7.10, 7.11 and 7.12).

Example analyses

TABLE 7.10 - Case 1: Water chemical analysis data sheet

Submitted by: Andreas Christoforou	**Date:** 11.9.97
Locality: Potamia	**Laboratory No.:** W-76/97
Analysis requested: Full ionic plus boron	**Borehole No.:** N332
Remarks: Planning cropping patterns - fruit trees, vegetables	
LABORATORY RESULTS	**Analyst:** N. Antoniou **Date:** 19.9.97
Electrical conductivity ECw dS/m: 3.6	**pH:** 7.1

Anions	mg/litre	meq/litre	Cations	mg/litre	meq/litre
Chloride (Cl^-)	429	2.09	**Sodium (Na^+)**	480	20.8
Sulphate (SO_4)	552	11.50	**Potassium (K^+)**	19	0.5
Carbonate (CO_3^{--})	Nil	Nil	**Calcium (Ca^{++})**	160	8.0
Bicarbonate (HCO_3^-)	480	7.90	**Magnesium (Mg^{++})**	60	5.0
Nitrate (NO_3^-)	180	2.90	**Boron (B)**	1.5	------
Total	1 641	34.3		719	34.3
TDS:	2 360				

Evaluation and remarks: SAR = 8, RSC = Nil
Medium saline water - High in sodium and boron content at toxic levels for most fruit trees (citrus, deciduous, etc.) grapes, strawberries and some vegetables (onion, garlic, beans) - there is no sodium hazard - under proper management, on light soil with good infiltration and internal drainage and no impermeable layer, it can be used for irrigation of crops tolerant to salinity and boron, such as olives, pomegranates, pistachio, date palms, most of the vegetables, watermelons, potatoes, etc. and forage crops - some delay in crop development and certain yield reduction should be expected - any problems from bicarbonates can be solved easily - due to high nitrate content, which is equal to 40 g of net nitrogen per cubic metre of water, the application of nitrogen fertilizer should be reduced by 66 percent for fruit trees and 20-30 percent for vegetables accordingly - frequent irrigation is recommended - LR 0.15.

Signature: A. Phocaides

TABLE 7.11 - Case 2: Water chemical analysis data sheet

Submitted by: N. Papas				**Date:** 2.10.1997	
Locality: Orini				**Laboratory No.:** W/400/97	
Analysis requested: Full ionic plus boron				**Borehole No.:** N335	
Remarks: Irrigation use					
LABORATORY RESULTS				**Analyst:** A.Magnetis **Date:** 9.10.1997	
Electrical conductivity ECw dS/m: 2.1				**pH:** 8.35	
Anions	**mg/litre**	**meq/litre**	**Cations**	**mg/litre**	**meq/litre**
Chloride (Cl^-)	215	6.05	**Sodium (Na^+)**	320	13.9
Sulphate (SO_4)	244	5.10	**Potassium (K^+)**	2	--
Carbonate (CO_3^{--})	Nil	Nil	**Calcium (Ca^{++})**	48	2.4
Bicarbonate (HCO_3^-)	432	7.11	**Magnesium (Mg^{++})**	31	2.6
Nitrate (NO_3^-)	41	0.66	**Boron (B)**	0.56	
Total	932	18.9		401	18.9
TDS:	1 333				

Evaluation and remarks: SAR = 9, RSC = 2.11
Slightly saline water - no sodium hazard or any severe toxicity problem - under proper management and on light soils with good structure and internal drainage it is suitable for use in the majority of the crops - bicarbonate may cause some micronutrient deficiency problems that can be overcome.

Signature: A. Phocaides

TABLE 7.12 - Case 3: Water chemical analysis data sheet

Submitted by:G. Demosthenous	**Date:**3.11.97
Locality: Limassol	**Laboratory No.:**
Analysis requested: Full ionic plus boron	**Borehole No.:**
Remarks: Irrigation of olives and other field crops	
LABORATORY RESULTS	**Analyst:** E. Iasonos **Date:**10.11.97
Electrical conductivity ECw dS/m: 2.3	**pH:**8.7

Anions	**mg/litre**	**meq/litre**	**Cations**	**mg/litre**	**meq/litre**
Chloride (Cl^-)	107	3.02	**Sodium (Na^+)**	410	17.80
Sulphate (SO_4)	278	5.80	**Potassium (K^+)**	6	0.10
Carbonate (CO_3^{--})	14	0.48	**Calcium (Ca^{++})**	8	0.40
Bicarbonate (HCO_3^-)	624	10.27	**Magnesium (Mg^{++})**	14	1.20
Nitrate (NO_3^-)	Nil	Nil	**Boron (B)**	2.88	
Total	1 023	19.5		438	19.5
TDS:	1 461				

Evaluation and remarks: SAR = 20, RSC = 8.67
Slightly saline water, however problematic - boron content at toxic levels for the majority of fruit trees and most herbaceous agricultural crops - danger of severe soil infiltration and permeability problem from the use of this water - excess bicarbonate salts could cause chlorosis to some plants - pH is higher than normal and may result in imbalanced nutrition - usage of this water should be done with caution, very good management, on light soils with high infiltration rate and permeability, and selected crops tolerant to boron toxicity, such as date palms, cabbage, cauliflower, squash, parsley, tomato, celery, asparagus, corn, alfalfa, sugar beet - the existing olive trees may be irrigated reservedly - soil improvement additives (washed manure, gypsum, etc.) should be applied occasionally - frequent irrigation is preferable - a follow-up based on a schedule is essential.

Signature: A. Phocaides

THE PHYSICAL QUALITY OF IRRIGATION WATERS AND TREATMENT (FILTRATION)

Irrigation water usually is not found in its pure state, but mostly with foreign solid particles and other impurities. The solid content in irrigation water mainly consist of dirt and suspended inorganic matter (*silt, sand, leaves, fine clay and rust dust*) and organic substances (*algae, bacteria, protozoa*) from vegetative origin and living organisms and bacteria populations. The introduction of improved irrigation systems with the use of closed pipes networks and small nozzles' passage water emitters, liable to blockages, necessitates the removal of the suspended solids to protect the emitters against clogging hazards.

The suspended solid content in the irrigation water may vary in a wide range and depends on the nature of the source. The four main sources of the irrigation water used with pressurized irrigation techniques and the primary kind of suspended solids (SS) content are:

a) Dams and open reservoirs: Green algae (phytoplankton), bacteria and zooplankton of different kinds and bacterial slime (sulphatic, iron and other), dissolved iron and manganese, and all kinds of other impurities from inorganic origin (debris, silt, clay etc.).
b) Underground water (wells and boreholes): Sand, silt, iron, manganese, sulphates and carbonates and bacteria.
c) Treated wastewater: Suspended solid particles of different sizes and shapes.
d) Water from pipe networks: Zooplankton (grown for years in the pipes in colonies up to 5 mm size and developed in 2–3 days in worms that blocked the filters).

Irrigation water quality cannot be clearly defined in respect of clogging problems, yet it can be primarily classified as Good – Moderate – Bad - Very Bad. The various substances in water, which contribute to the clogging of the micro-irrigation systems can be divided into three main categories of:

- Suspended particles of organic and inorganic matter,
- Precipitate forming elements (iron, manganese, calcium, magnesium),
- Bacterial slimes.

For an in-depth evaluation it is necessary to examine the Total Suspended Solids, the Particle size Distribution, the Total Dissolved Solids, the water pH, the Hardness, the Turbidity, Iron and Manganese, the Hydrogen Sulphide and the Microbial population.

Filtration

The filtration of the irrigation water is essential in order to avoid blockages damage of the small passage emitters, drippers, sprayers and sprinklers. This is a mechanical water treatment and it is achieved by the installation of main filtering devices (filters) at the Control head of the irrigation systems. It is an integral part of the pressurized systems installations. The size of flow (capacity of filtration) may affect the type of the filtration in major water works, but for low rate applications at the farm level the type and the degree of filtration depends on:

- The kind of suspended matter in the irrigation water, and
- The filtration requirement of the system (emitters).

Most of the filters available for irrigation water are:

- the **gravel or granular (sand media) filters**, which operate on the principle of in-depth filtration and effectively withhold the large particles of unbroken organic mutter (algae) and dust,
- the **hydrocyclones (sand separators)**operate on the vortex flow principle and are used to collect large quantities of sand contained in pump underground water,
- the **strainers (screen type or disk/grooved rings type)** effective in filtering the inorganic suspended matter. These are equipped with filtering elements with perforations smaller than the emitters' diode (up to 70 percent).

In cases where water contains all kind of suspended matter it is necessary to install all three types of filters. The Hydrocyclone and the Gravel filter are always placed at the upstream of the Control Head of the System and the Strainer filter at the downstream. The latter is always installed in every micro-irrigation system after the fertilizer apparatus. The degree of filtration is usually given in the old English unit "mesh" (number of perforations per linear inch) or in "microns". The filtration requirements of various water emitters are:

- 16–25 mesh (1000–500 micron) for Slow-rotation impact drive sprinklers medium pressure
- 60–100 mesh (250–120 micron) for mini and micro-sprinklers and sprayers.
- 80–160 mesh (200–100 micron) for dripper emitters.

Operation and maintenance

Proper operation and maintenance (cleaning) of the filters is of paramount importance for effective filtration and to avoid the build up of a "filter cake" generating further clogging problems to the systems. Some filters are made for complete automatic and unattended operation, others are automatic self-cleaning and others are of self-cleaning or manual flushing and cleaning mechanism (a brief description of the filters is given in Chapter 3). The systems hydraulic pressure operates the filters, but automation requires electricity (AC or DC). The frequency of cleaning is planned either at fixed time intervals between two consecutive cleanings or whenever the differential pressure along the filter increases to the initial normal value (0.2–0.3 Bars). The suppliers give detail instructions for back flushing (size and velocity of flow, stream direction etc.) according to the filter mechanism and mode of operation.

Application of chemicals

Application of chemicals into the irrigation water, before the filtration system, may reduce the quantity of suspended matter, control the bacteria growth in the system network, decompose algae and dissolve the solid particles. It also prevents sedimentation. Copper sulphate is used extensively in reservoirs at a maximum concentration of 2 ppm to control growth of algae. Treatment with acid reduces the pH of the water thus prevents precipitation of dissolved solids and dissolves the existing precipitations. It can prevent carbonate precipitation too. The various acids recommended are the Hydrochloric acid (HCl) the Sulphuric acid (H_2SO_4) and Phosphoric acid (H_3PO_4). The phosphoric acid is a fertilizer too, however the concentration should be high enough to decrease pH < 6 and prevent phosphorous sediments. The quantity of acid depends on the requested pH of the water.

The safest and less expensive chemical for use with irrigation water, at normal pH and temperatures around 20°C, is chlorine in the form of Sodium Hypochlorite (NaOCl). Contact time also influences the effectiveness of chlorination. It is available everywhere as household liquid at concentrations of 2–15 percent available chlorine. The application is done through continuous or intermittent injection during irrigation at low uniform concentrations of approx. 5 ppm and 10 ppm respectively. Control of effective chlorination is achieved by the measurement of the free residual chlorine concentration in the water. This should be around 1.0–2.0 ppm at the end of the irrigating pipe. Use of ammonia fertilizer should be avoided during chlorination. Over dosage of chlorine into the systems networks may result into the movement of sediments cause severe clogging of the emitters.

QUALITY OF TREATED WASTEWATER FOR IRRIGATION (PHYSICAL, BIOLOGICAL AND CHEMICAL)

The treated wastewaters are a new source of water, which is expected to cover gradually more than 10 percent of the water requirements for agricultural and landscape irrigation. Treated wastewater may possess various chemical contaminants (salts, nutrients and trace elements) and biological undesirable constituents (water born pathogens, i.e. helminths, protozoa, bacteria and viruses shed in the excreta of healthy people and ill persons). Uncontrolled use of this type of water frequently is associated with significant negative impacts on human health and the environment. These impacts can be minimized when good management practices are implemented. So, it introduces a new element regarding the water quality evaluation for irrigation. "Wastewater" refers to domestic sewage and municipal wastewaters that do not contain substantial quantities of industrial effluents.

Evaluation criteria and parameters

The use of reclaimed wastewater is always planned, designed and managed properly. If not it might become hazardous for the people, the livestock and the environment. World Health Organization (WHO) published Guidelines for the Safe Use of Wastewater and Excreta in Agriculture and Aquaculture in 1989. These guidelines are currently under revision with expected publication in 2004. In this context some additional water quality criteria are considered based on water suitability for re-use in agriculture and landscape irrigation, and on ways and methods of improvement and management to satisfy the requirements of the installations, the operators and the yield consumers.

The evaluation of the treated effluents is based on world-wide established criteria, which refer to limiting values of certain physical, chemical and biological parameters in order to avoid possible adverse responses, when used, or disposed. The following parameters, when considered can give a correct picture of the usability of the treated wastewater and the level of its suitability for irrigation purposes:

Chemical Parameters:
- Total Salinity, ECw dS/m, TDS mg/l,
- Acidity/Basicity, pH
- Hardness, $CaCO_3$ mg/l
- Types and concentration of anions and cations, me/l
- Sodium adsorption Ratio, SAR
- Nitrate – Nitrogen, NO_3–N mg/l
- Phosphate Phosphorus, PO_4–P mg/l
- Trace Elements, mg/l
- Heavy Metals, mg/l

Physical and Biological Parameters:
- Colour,
- Odor,
- Turbidity NTU,
- BOD 5 mg/l (Biochemical Oxygen Demand)

- COD mg/l (Chemical Oxygen Demand)
- SS mg/l (Suspended Solids)
- Total Coliforms/100 mg
- Faecal Coliforms
- Intestinal Nematodes

Impact on soils and plants

The evaluation of the treated wastewater as it is concerned with its chemical quality is well understood, as it has already been described previously. It is made for any possible long-term effect on irrigated soils and plants and groundwater protection. Experience gained so far, suggests that with proper management (improved irrigation system, proper irrigation schedule) salinity/toxicity and nitrogen effects on soil can be under control. Excess NO3 may cause some problems to the plants. The trace elements and heavy metals, which are present in many treated effluents rarely, exhibit serious toxicities to plants although in most cases they accumulate in the plants themselves. When the plants are consumed by livestock possible health hazards can developed. E.g. lead and vanadium is toxic to forage crops at normal and low concentrations respectively. Still the time required for development of health hazards to the livestock is very long, 20–50 years and if repeated applications of heavy metals in excess of the maximum permissible are practiced. Even then it does not mean that phytotoxicity will occur. No problem has been observed with the accumulation of heavy metals in the crops or in the soil. In fact the conventional fertilizers were found to add far greater quantities of heavy metals.

Effects on the environment

As it is concerned with the effects on the environment, the values of the chemical and biological parameters for the treated effluents should be at levels acceptable for the purpose of use and disposal. There is always the risk that wastewater irrigation may facilitate the transmission of intestinal nematode infections and faecal bacterial diseases to both consumers and agricultural workers. In fact there is no practical experience and knowledge on the wastewater pollutants movement downwards and the current devices and sensors detecting the contaminants within the soil are too expensive. However, it is known that nitrates contained in the irrigation water are mobile and eventually reach the groundwater. There is no doubt that the same happens with boron. As regards with the toxic heavy metals, studies have indicated that more than 85 percent of the applied trace elements accumulate in the surface few centimetres of the soil. Yet it is possible that long term application of treated effluents with toxic elements, especially under acid soil conditions and on sandy soils might lead ultimately to their mobilization and leaching down in deeper layers and finally to pollute the groundwater. It must be underlined that no such cases have been reported until recently.

Effects on the irrigation system

Treated wastewaters contain generally large quantities of suspended organic matter, other waterborne trash and impurities of inorganic origin that cause blockages and clogging of emitters on the irrigation systems.

Health protection (after WHO)

From the standpoint of public health protection, the available measures can be grouped under the following four main categories:

- ***Treatment process and degree of pathogens removal:*** Removal of pathogens is the prime objective in treatment of wastewaters for reuse. Various conventional processes for primary and secondary treatments, **plain sedimentation, activated sludge, biofiltration (trickling filters), aerated lagoons, oxidation ditches, row-sewage disinfections and waste stabilization ponds** cannot remove the bacterial and helminth egg effectively. **Supplementary disinfections and filtration**(tertiary treatment) is needed to produce quality water at accepted levels to comply with recommendations for unrestricted irrigation. In arid and semi-arid regions the Waste stabilization ponds are used. With a minimum retention time of 11 days and depending on the temperature with about twice that time this treatment process may achieve adequate pathogens removal (helminth and bacteria) to meet the recommended standards for reuse. Additional "polishing" ponds are needed to the conventional treatment plants.

- ***Crop selection:*** Based on the treated water quality the various crops to be grown without risk to the consumer are categorized (A, B and C) in accordance with the required extent of measures for health protection. (WHO recommended microbiological quality guidelines for wastewater use in agriculture, 1989)?

  - Category A - protection for field workers includes industrial crops, grains and forestry and food crops for canning.
  - Category B - protection for consumers, farmers and the general public, applies to pasture, green fodder and tree crops, as well as to fruit and vegetables that are peeled or cooked before eating.
  - Category C - unrestricted irrigation, covers fresh vegetables, spray irrigated fruit, grass and lawns in public parks, sport fields etc.

  It must be underlined that the crop selection in category B provides protection only to consumers and not to the farmers and field workers. Additional measures should be applied, such as control of human exposure.

- ***Method of irrigation application:***
  Irrigation with pressurized irrigation systems has many advantages over the traditional surface methods. Gravity surface methods expose the farmers to the greatest risk. This is eliminated with the use of Pipes Distribution surface systems. However if the treated water is not recommended for unrestricted irrigation, sprinkling and spraying methods should not be used for crops likely to be eaten uncooked, fruits and for grass and lawns in public parks, sport fields etc. Drip irrigation (conventional) and bubblers can give a great degree of health protection. Drip irrigation with mulching and sub-surface gives the highest protection.

- ***Human exposure:***
  The people at potential risk from the use of treated wastewater are the farmers and their families, the crop handlers, the consumers of products and those living near the effected fields. The risk can be reduced by several precaution measures such as are the proper management, the immunization against typhoid and hepatitis and other infections, the limited exposure by the use of the right clothe and footwear, thorough cooking of food, and other hygiene measures.

National guidelines and standards (The Cyprus case)

The national guidelines for reuse of treated wastewater for irrigation in many countries mostly comply with those published by WHO. The objectives of WHO is to provide background and guidance to governments for developing their national standards with respect to international health matters, the protection of the public health and the preservation of the environment. An excellent example is the "Cyprus case". In this country the reuse of treated effluents is a relatively new practice, it started in mid-eighties, however a lot of progress has been made through and now it can be used as a model for many other countries.

The Cyprus Standards

The Sewage Effluent and Sludge Technical Committee under the Ministry of Agriculture, Natural Res. and Environment of the Republic of Cyprus prepared quality standards for the treated effluents and sludge to be reused in agriculture and amenity areas. The standards are composed of two parts, *the Guidelines and the Code of Practice* and at the moment they seem to be the most advanced as compared with other national standards, so they are used in this chapter.

The *"Guidelines"* includes the variables of BOD as the main indicator for better monitoring and control of the treatment process, Suspended

Solids for the effective disinfections against pathogens and especially viruses, as well as for the elimination of filters clogging and micro-emitters blockages. Faecal Coliforms are included as an indicator of pollution and Intestinal Nematodes as the parameter directly referring to a group of pathogens and as an indicator for protozoa removed from the treated wastewater as intestinal warms. The methods of treatment (secondary, tertiary/storage and disinfections or stabilization maturation) are also included to ensure the quality parameters. The *"Code of Practice"* is complementary regulations yet inseparable part of the *"Guidelines"* specifying treatment techniques, irrigation methods for each crop and criteria for handling the systems of irrigation, security measures, etc. It incorporates additional measures, which act as barriers to the transmission of diseases. The techniques/methods for tertiary treatment are also specified.

Pressurized irrigation with treated wastewater

All pressurized irrigation systems described in this Handbook can be used for irrigation with treated wastewater. The water microbiological quality standards, the farming practice, and the kind of crop are the main interrelated factors that affect the selection of the irrigation system. Different techniques and methods are suggested for the irrigation of different crops for many reasons.

The micro-irrigation systems in general (drip, low-capacity sprinklers, bubbler and mini-sprinkler) are suitable for irrigation with this type of water, as these permanent localized methods secure for the minimum contact of treated water with crops and farmers (Figures 7.2 and 7.3). Conventional sprinkling (solid installations) and mechanized spraying systems (center pivot and traveller boom) are conditionally suitable. Low-cost systems (hose-basin, hose furrow and pipe distribution surface systems) are also recommended. Their use mainly depend on the quality of the effluent, the kind of crop, the potential risk to the health of the workers, the public and the environment and finally on the background and skill of the farmers to handle this type of water. For further details please read the *"Cyprus Code of Practice"*(see Table 7.13), which provides adequate and specific information for nearly all crops and the appropriate irrigation techniques. The installation of efficient filtration at the Control Head of the systems is of major importance as it is the arrangement for automation in sprinkling and spraying methods of water application.

FIGURE 7.2 - Drip maize with recycled water.

FIGURE 7.3 - Sprinkling with treated municipal water.

TABLE 7.13 - Cyprus Guidelines for Domestic Treated Effluents use for Irrigation

Irrigation of:		BOD mg/l	SS mg/l	Faecal Coliforms /100 ml	Intestinal Worms/l	Treatment required
All crops (a)	(A)	10*	10*	5* 15**	Nil	Secondary and Tertiary and Disinfection
Amenity areas of unlimited access. Vegetables eaten cooked (b)	(A)	10* 15**	10* 15**	50* 100**	Nil	Secondary and Tertiary and Disinfection
Crops for human consumption. Amenity areas of limited access	(A)	20*	30* 45**	200* 1 000**	Nil	Secondary and Storage > 7 days and Disinfection or Tertiary and Disinfection
	(B)	--	--	200* 1 000**	Nil	Stabilization-maturation ponds total retention time > 30 days or Secondary and storage> 30 days
Fodder crops	(A)	20* 30**	30* 45**	1 000* 5 000**	Nil	Secondary and Storage > 7 days and Disinfection or Tertiary and Disinfection
	(B)	--	--	5 000*	Nil	Stabilization-maturation ponds total retention time > 30 days or Secondary and storage> 30 days
Industrial crops	(A)	50* 70**	--	3 000* 10 000**	--	Secondary and Disinfection
	(B)	--	--	3 000* 10 000**	--	Stabilization-maturation ponds total retention time > 30 days or Secondary and storage> 30 days

(A) Mechanized methods of treatment (activated sludge,
(B) Physical methods of treatment (Stabilization Ponds)
* These values must not be exceeded in 80 percent of samples per month. Min. No of samples 5
** Maximum value allowed
(a) Irrigation of leafy vegetables, bulbs and corms eaten uncooked is not allowed
(b) Potatoes, beet-roots, colocasia

Note:
No substances accumulating in the edible parts of crops and proved to be toxic to humans or animals are allowed in effluent.

Code of Practice for Treated Sewage Effluent used for Irrigation in Cyprus

1. This sewage treatment and disinfection must be kept and maintained continuously in satisfactory and effective operation so long as treated sewage effluent are intended for irrigation, and according to the license issued under the existing legislation.

2. Skilled operators should be employed to attend the treatment and disinfection plant, following formal approval by the appropriate authority that the persons are competent to perform the required duties, necessary too ensure that conditions (1) are satisfied.

3. The treatment and disinfection plant must be attended every day according to the program issued by the Authority and records to be kept on all operations performed according to the instructions of the appropriate Authority. A copy must be kept for easy access within the treatment facilities.

4. All outlets, taps and valves in the irrigation system must be secured to prevent their use by unauthorised persons. All such outlets must be coloured red and clearly labelled so, as to warn the public that water is unsafe for drinking.

5. No cross connections with any pipeline or works conveying potable water, is allowed. All pipelines conveying sewage effluent must be satisfactory marked with red so as to distinguish them from domestic water supply. In unavoidable cases where sewage/effluent and domestic water supply pipelines must be laid close to each other the sewage or effluent pipes should be buried at least 0.5 m below the domestic water pipes.

6. Irrigation methods allowed and conditions of application differ between different plantations as follows:

   6.1. Park lawns and ornamental in amenity areas of unlimited access:
   - Subsurface irrigation methods
   - Drip irrigation
   - Pop-up sprinklers, low-pressure, high precipitation rate lo-angle < 11°. Sprinkling preferably practiced at night and when people are not around.

   6.2. Park lawns and ornamental in amenity areas of limited access, industrial and fodder crops
   - Subsurface irrigation
   - Bubblers

- Drip irrigation
- Surface irrigation methods
- Low capacity sprinklers
- Spray or sprinkler irrigation, is allowed with a buffer zone of about 300 meters.

For fodder crops irrigation is recommended to stop at least one week before harvesting and no milking animals should be allowed to graze on pastures irrigated with this sewage. Veterinary services should be informed.

6.3. Vines:
- Drip irrigation
- Mini-sprinkler and sprinklers (in case where crops get wetted, irrigation should stop for two weeks before harvesting).

Moveable irrigation systems not allowed No fruit should be selected from the ground.

6.4. Fruit trees
- Drip irrigation
- Hose-basin irrigation
- Bubbler irrigation
- Mini-sprinklers irrigation

No fruit to be collected from the ground, except for nut-trees. In case where crops get wetted, irrigation should stop one week before harvesting.

6.5. Vegetables
- Subsurface irrigation
- Drip irrigation

Crops must not come in contact with effluents.
Other irrigation methods could also be considered.

6.6. Vegetables eaten cooked
- Sprinklers
- Subsurface irrigation
- Drip irrigation

Other irrigation methods may be allowed after the approval of the appropriate Authority. Restrictions may be posed to any method of irrigation by the appropriate Authority in order to protect public health or environment.

7. The following tertiary treatment methods are acceptable:

    7.1. Coagulation plus flocculation followed by rapid sand filtration
    7.2. Slow Sand Filters
    7.3. Any other method, which may secure the total removal of helminth ova and reduce faecal coliforms to acceptable levels, must be approved by the appropriate Authority.

8. Appropriate disinfection methods should be applied when sewage effluents are to be used for irrigation. In the case of chlorination the total level of free chlorine in the effluent at the outlet of the chlorination tank, after one hour of contact time should be at least 0.5 mg/l and not greater than 2.0 mg/l.

9. Suitable facilities for monitoring of the essential quality parameters should be kept on site of treatment.

CHAPTER 8: Hose-move sprinkler irrigation

INTRODUCTION

In recent decades various sprinkler irrigation methods and installations, both solid and portable, have been developed to meet farmers' needs. The most widely adopted and least expensive system for irrigating small to medium-sized farms is the piped hand-move system with a low to medium operating pressure (2.0–3.5 bars). The sprinklers are mounted at equal spacings (6–12 m) on the lateral pipelines laid across the field at predetermined intervals (called lateral positions) of 6–18 m so that the irrigation water is sprinkled uniformly over the area covered (Figure 8.1).

To avoid lateral movement and to reduce labour requirements, the hose-move sprinkler system has been developed. It is an improvement on the conventional piped hand-move system and combines some features of semi-permanent installations with those of permanent ones. In this system the sprinkler lateral lines are placed permanently at a wide spacing up to 60 m apart. The sprinklers, mounted on tripod stands, are not fitted directly to the lateral pipes, but connected to them via flexible PE hoses which are 20–25 mm in diameter and up to 30 m in length. The hoses with the sprinklers can be moved laterally on either side to cover a number of lateral positions.

As the sprinklers are of low to medium pressure, this system can be classed as a low or medium pressure, semi-permanent, hand-move installation. It is recommended for the irrigation of full coverage crops such as alfalfa, maize, cotton, potatoes, carrots and groundnuts. It should be noted that hose-move systems are different from the drag-hose system. The latter is used only for under-tree sprinkling and the sprinklers are fitted on small skids which can be easily dragged backwards from a distance.

FIGURE 8.1 - Hose-move sprinkler irrigation.

SYSTEM LAYOUT AND COMPONENT PARTS

The layout of the system is the standard one consisting of a head control, a pipe distribution network (mains, submains and manifolds, where needed), hydrants, laterals and a number of hoses (one per sprinkler).

The head control is simple and includes only the regulating valves (shutoff, non-return, air, etc.). The main and submain pipelines are usually buried 90–150 mm rigid PVC pipes, or 75–110 mm HDPE pipes laid on the surface. The hydrants (2–3 inches) are located along the manifolds (mains or submains) at the same wide spacing as the sprinkler laterals. The manifolds and the sprinkler laterals can be either HDPE or quick coupling light steel/aluminium pipes (63–75 mm). The flexible hoses are soft 20–25 mm LDPE pipes. The tripod stands for the sprinklers can be made of 8 mm iron rods.

SPRINKLERS

The water discharged through the sprinkler devices is shot into the air and falls to the ground in a circular pattern around the sprinkler. Most of the agricultural sprinklers have a hammer-drive slow-rotating or revolving mechanism (hammer wedge and spring, or hammer and rocker weight) and use a low to medium operating pressure (2.0–3.5 bars). They are equipped with two nozzles for discharging the water: the range and the spreader. The range nozzle, larger in diameter, shoots a water jet and covers the area distant from the sprinkler, activating the rotating mechanism at the same time. The spreader nozzle sprays the water in the vicinity of the sprinkler. The nozzles are interchangeable to allow variability in performance according to requirements. The sprinklers are made of brass or heavy-duty plastic. Most of them have several parts made of brass and others of plastic. The axle and the spring are made of stainless steel. The main performance characteristics of the sprinklers used in hosemove systems are:

- two nozzles: 3.0–6.0 mm (range) *x* 2.5–4.2 mm (spreader);
- low to medium operating pressure: 1.8–3.5 bars;
- water discharge: 1.1–3.0 m^3/h;
- diameter coverage (wetted): 18–35 m;
- jet angle: 20°–30° (except where low angle jet is needed, e.g. strong wind, treated water);
- type of connection: threaded internal or external ½–1 inch.

To ensure satisfactory sprinkling with impact rotating conventional sprinklers, the minimum operating pressure should be at least 2.0 bars.

DESIGN CRITERIA AND CONSIDERATIONS

The water discharged from a single sprinkler is not uniformly distributed over the entire area; a greater quantity falls near the sprinkler and less in the periphery. To ensure a uniform precipitation over the entire area under irrigation, the sprinklers are always placed so that they overlap each other from both directions. This setting is termed sprinkler spacing. The spacing of the sprinklers along the lateral lines is known as SL, and the spacing between two lines as Sm. The spacing pattern is square, rectangular or triangular, with SL = Sm.

In order to obtain good distribution uniformity by overlapping, the sprinkler spacing (Sm) should not exceed 65 percent of the sprinkler diameter coverage under light to moderate wind conditions in the square and rectangular patterns. In the triangular pattern, the spacing can be extended up to 70 percent of the diameter coverage. In strong wind conditions, the spacing should be 50 percent of the diameter coverage with the lateral direction perpendicular to the wind direction. With a wind strength of over 3.5 m/s, sprinkling is not recommended (Figure 8.2).

The average application (precipitation) rate is a function of the sprinkler discharge and the spacing of the sprinklers:

Precipitation rate (mm/h) = sprinkler discharge (litres/h) ÷ SL *x* Sm (m)

The rate of precipitation should not exceed the soil intake (infiltration) rate (25 mm/h in light soils, 8–16 mm/h in loams and 2–8 mm/h in clays).

FIGURE 8.2 - Profile of wetted soil under sprinklers

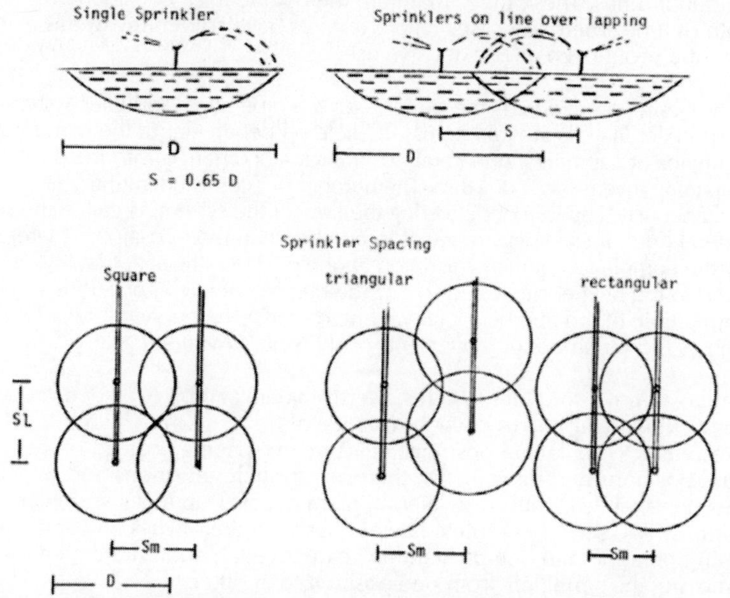

D = Diameter Coverage, SL = spacing along the line, Sm = spacing between the lines. Under normal wind conditions (wind speed 2m/sec) S=0.60D. In strong wind (2.5-3.5 m/sec) S=0.50D.

WIND CONDITIONS

Wind directions and velocities must be recorded and classified accordingly, (0–0.7 m/sec Nil wind, 0.7–2.5 m/sec Light, 2.5–3.5 m/sec Moderate to Strong, and > 3.5 m/sec very Strong). Sprinkling is not recommended under strong wind conditions.

The common sprinkler spacing in low-medium pressure systems is 6, 9, or12 m along the laterals and 12 or 18 m between the laterals. Initially, these spacings were convenient given the standard length of the quick coupling pipe (6 m). However, they have proved most practical as the close spacing, low discharge and precipitation rates of 8–14 mm/h give better results. The height of the sprinklers above ground should be a minimum of 60 cm for low-growing crops. For high-growing crops, the height should be adjusted accordingly.

The light portable quick coupling pipes (steel or aluminium) can be used not only as sprinkler lateral lines but also as water conveyance and distribution lines. These pipes maintain their value for a considerable length of time. There are cases where farmers have sold many of these pipes at a profit even after extensive use.

The design procedure is the same as for the pipe-move sprinkler systems. The sprinkler laterals are laid across the field perpendicular to the manifold line (mains or submains) on lateral positions in accordance with the designed Sm spacing, every 6, 12 or 18 m. The number of laterals operating simultaneously, capable of delivering the flow of the system, is called the set of lateral lines; these lines are fewer in number than their positions. Therefore, after the completion of their operation at one position, the set of laterals is moved to the next position and so on. The number of lateral positions should be a multiple of the number of lateral lines per set. The quotient of the two numbers is the number of movements or shifts per irrigation cycle.

In hose-move sprinkler systems, the sprinklers can be extended on both sides of the lateral line to cover a distance of up to 60 m, which corresponds to six lateral positions at 12 m Sm spacing. Instead of the lateral positions and movements, there are sprinkler positions and movements (shifts). Thus, one lateral line may cover up to six sprinkler positions. Two sets of complete lateral lines with their hoses and sprinklers, one in operation and one on stand-by, can cover an entire field area just by moving the sprinklers from one position to another.

The maximum permissible length of the lateral lines is a function of the size of the pipe, the number of the sprinklers (the spacing) and their discharge. The loss of pressure due to friction in the lateral line should not exceed 20 percent of the pressure at its entrance. Based on this assumption, some indicative figures for quick coupling light steel or aluminium laterals are presented in Table 8.1.

TABLE 8.1 - Maximum number of low-medium pressure sprinklers on quick coupling lateral pipes

Sprinkler pressure bars	Sprinkler discharge m³/h	50 mm		70 mm		89 mm	
		SL spacing					
		6 m	12 m	6 m	12 m	6 m	12 m
2.5 3.0 3.5	1.5 1.65 1.8	12	10	23	18	36	28
2.5 3.0 3.5	2.0 2.2 2.3	10	8	19	15	30	23

IRRIGATION SCHEDULING PROGRAMME

With sprinkler irrigation, the whole area is wetted and, thus, a larger volume of soil is wetted. This allows a relatively higher water content in the soil to be maintained than is the case with localized methods, thereby increasing the irrigation interval. The larger the volume of wetted soil, the later the crop goes into deficit. The preparation of the irrigation programme follows the standard procedure, i.e. taking into consideration the soil moisture holding capacity, the plant physiology (root depth, growing stages, crop coefficient, etc.) and the climate. The irrigation efficiency is about 75 percent. In general, the irrigation dosage application depth for deep rooted field crops under sprinkling ranges from 40 to 100 mm. With a precipitation rate of about 14 mm/h, the operating time at each position is approximately 3 to 7 hours. Irrigation intervals of two weeks are common in sprinkler irrigation.

COST

The total cost for the installation of the system in 2.0 ha (as in the example design) is US$1 790, or less than US$1 000/ha. A cost analysis shows that the head control costs about US$70. The major cost items are the plastic pipes, PVC and PE tubes, for the system's network which amount to US$1 177, 65.7 percent of the total cost. Imported sophisticated equipment, such as the sprinklers, rarely exceeds 10 percent of the total cost.

ADVANTAGES

- High irrigation application efficiency – 75 percent.
- Easy design, simple installation and operation.
- Adaptability for all types of soils, many kinds of field crops and small irregular plots.
- Less expensive than many other modern irrigation systems.
- Involves unskilled labour.

DISADVANTAGES

- Moving the hoses with the sprinklers is heavy and unpleasant work (Figure 8.3).
- Long duration for the irrigation cycle.

FIGURE 8.3 - Hose move sprinkler irrigation (Sample design) 2.0 ha

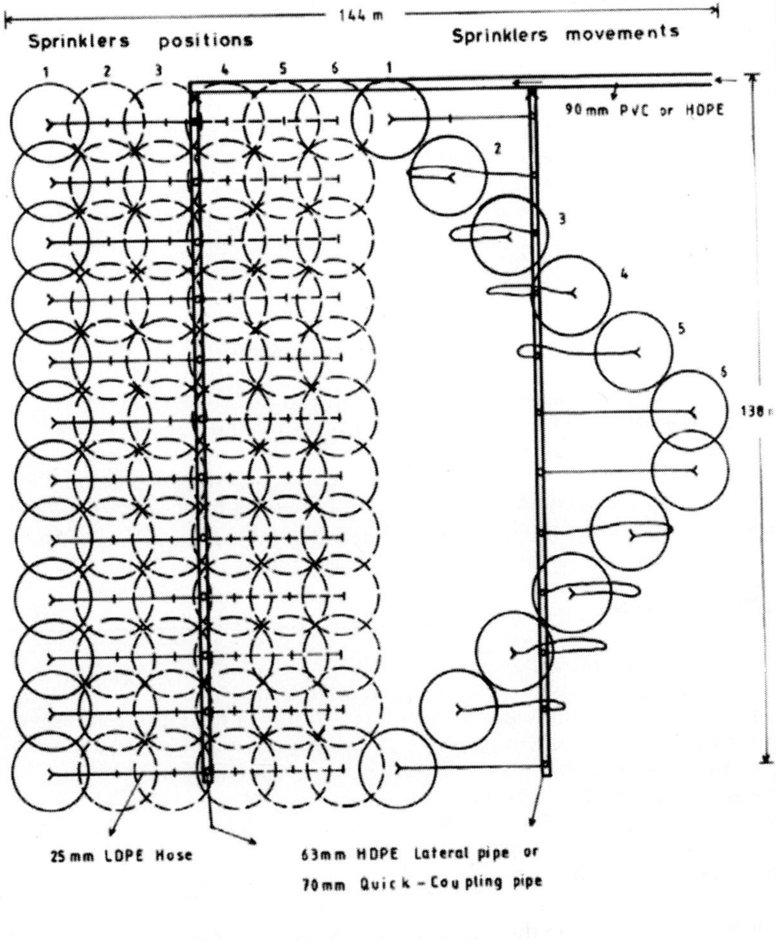

EXAMPLE DESIGN – Hose-move sprinkler for cotton

Area and crop

An area of approximately 2.0 ha planted with cotton at the beginning of August. The field is square and level.

Soil, water and climate

Medium texture soil of good structure, with good infiltration and internal drainage. The soil available moisture (Sa) is 110 mm/m depth. The water is of good quality with no salinity or toxicity hazards; the source is a tube-well equipped with a pumping unit delivering 36 m^3/h. The peak irrigation demand is in October, at the midseason growth stage of the crop.

Crop water requirements and irrigation scheduling

The pan average readings in October are 5.6 mm/d. This figure multiplied by 0.66 (pan correction factor) gives an ETo of 3.7 mm/d. The crop factor kc for cotton at this stage is taken as 1.05, the root depth 1.0 m and the moisture depletion 50 percent. Then, ETc cotton = 3.7 *x* 1.05 = 3.88 mm/d. The net application depth is Sa 110 mm *x* root depth 1.0 m *x* depletion 0.5 = 55 mm. The maximum permissible irrigation interval in October is 55 mm ÷ 3.88 mm/d = 14 days. The irrigation frequency depends on many factors and in no case should exceed the maximum permissible irrigation interval. The system's application efficiency is 75 percent, therefore, the gross application depth at peak is: 55 mm ÷ 0.75 = 73.3 mm. The gross irrigation dose is: 73.3 mm *x* 10 *x* 2.0 ha = 1 466 m^3.

System layout, performance and hydraulics

A 90 mm rigid PVC main pipeline is buried along the northern boundary of the field. Two 63 mm HDPE lateral pipelines are placed perpendicular to the mains, from north to south, 60 m apart and connected with the mains through offtake surface hydrants. On the lateral lines and at a regular spacing of 12 m, 25 mm flexible PE hoses 30 m long are fitted and extended on the sides. At the other end of the hoses are sprinklers on tripod stands (Tables 8.2 and 8.3 and Figure 8.4).

- sprinkler characteristics and performance: low pressure, two-nozzle sprinklers, discharge = 1.5 m^3/h at 2.5 bars operating pressure, diameter coverage = 26 m;
- sprinkler spacing: 12 *x* 12 m;
- precipitation rate: 10.4 mm/h;

- number of sprinklers per lateral: 12;
- number of laterals: 2;
- total number of sprinklers: 24 (operating simultaneously);
- lateral discharge: 18 m^3/h;
- system discharge: 36 m^3/h;
- number of (lateral) sprinklers positions (shifts): 6;
- duration of application per shift: 73.3 mm ÷ 10.4 = 7 hours;
- duration of irrigation cycle: 42 hours.

TABLE 8.2 - Total dynamic head required

	bars
Pressure required at sprinkler	2.50
Friction losses in the 25 mm LDPE sprinkler hose, 30 m	0.33
Friction losses along the 63 mm HDPE lateral line	0.47
Friction losses along the 90 mm PVC main line	0.15
Minor local and other losses	**0.25**
Total dynamic head required	3.70

TABLE 8.3 - List of the equipment required for the hose-move sprinkler system installation (bill of quantities)

Item	Description	Quantity	Unit price US$	Total price US$
	System distribution network			
1.	90 mm rigid PVC pipe, 6 bars	110 m	2.50	275.00
2.	63 mm HDPE pipe, 6 bars	280 m	1.80	504.00
3.	3-in *x* 90 mm PP adaptor	1 pc	10.00	10.00
4.	2 ½-in *x* 63 mm PP adaptor	2 pcs	5.00	10.00
5.	90 mm PP end plug	1 pc	10.00	10.00
6.	63 mm PP end plug	2 pcs	5.00	10.00
7.	90 mm *x* 2 ½ in PP clamp saddle	2 pcs	3.00	6.00
8.	63 mm *x* ¾ in PP clamp saddle	24 pcs	1.30	31.20
9.	¾ in *x* 25 mm PP adaptor	48 pcs	1.00	48.00
10.	2 ½ in threaded riser pipe, 60 cm	2 pcs	4.00	8.00
11.	2 ½ in gate valve	2 pcs	13.00	26.00
12.	2 ½ in nipple	2 pcs	1.00	2.00
13.	Tripod sprinkler stand	24 pcs	8.00	192.00
14.	Sprinkler two nozzle, 1.5 m^3/h at 2.5 bars	24 pcs	8.00	192.00
15.	25 mm LDPE hose, 4 bars	720 m	0.40	288.00
	Trench excavation and backfilling	110 m	1.00	110.00
	Sub-total			**1722.20**
	Head control			
16.	2 ½ in brass check valve	1 pc	15.00	15.00
17.	2 ½ in brass shut-off valve	2 pcs	13.00	26.00
18.	2 ½ in tee (galvanized iron, or PVC)	3 pcs	3.50	10.50
19.	2 ½ in nipple	4 pcs	1.00	4.00
20.	1 in single air valve	1 pc	12.00	12.00
	Sub-total			**67.50**
	TOTAL COST			**1789.70**

FIGURE 8.4 - Hose move sprinkler jointing techniques

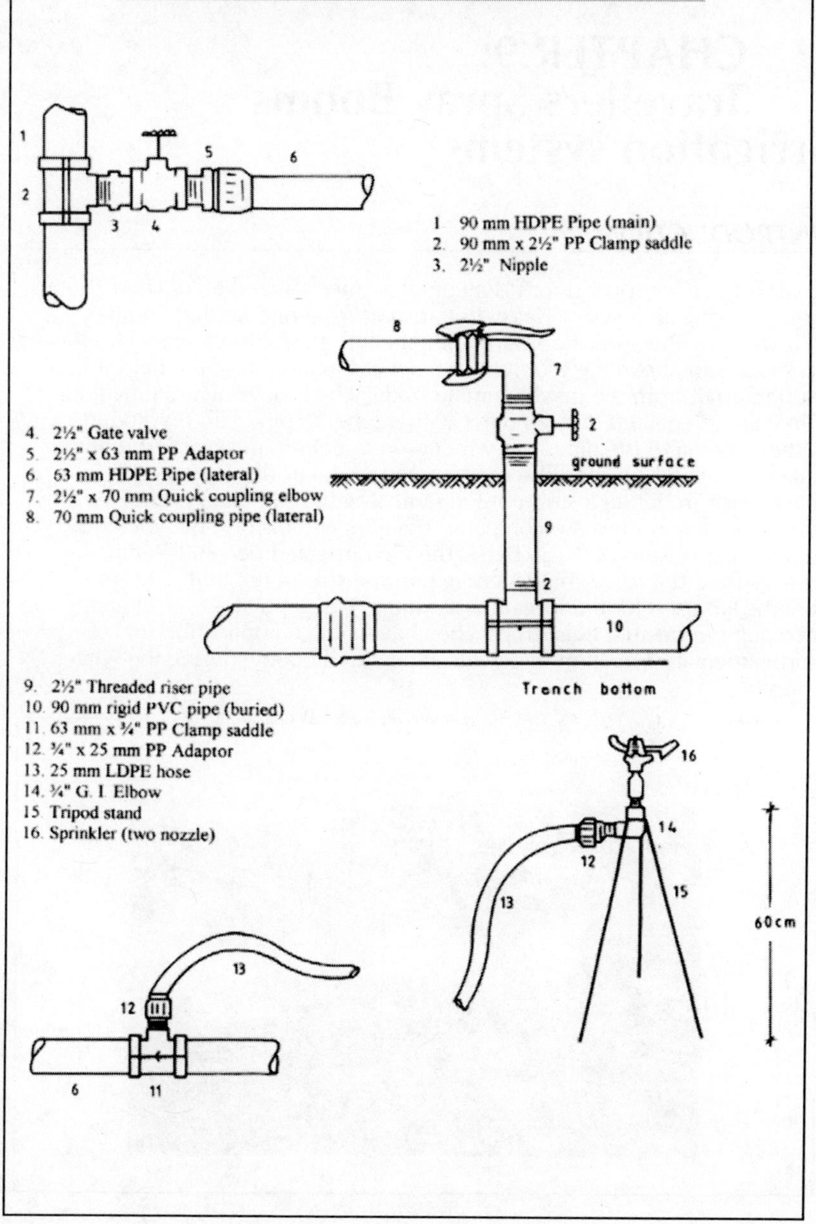

CHAPTER 9: Travellers Spray Booms irrigation systems

INTRODUCTION

The Traveller Spray Boom is a complete mechanized automated spraying irrigation system, easy to transport from one field to another (Figure 9.1). The spray boom-irrigating lateral, placed on a wheel trolley 1.3 to 2.5 m above the ground, is towed at the far end of the field up to 400 m apart from the machine main body, which stays near the hydrant. They are connected to each other with a long PE pipe laid on ground. When operated, the pipe is re-winded on a reel (dram) attached to the machine main body and the spray boom on the trolley is dragged backwards irrigating a strip until the full length of the field is covered. The traveller spray booms are compact irrigation systems of low – medium operating pressure (3.0–4.5 Bars). The area irrigated per shift (setting) ranges from 0.4 to 2.0 ha depending on the size of the unit. The Spray booms are used for the irrigation of forage, grain, potatoes, groundnuts and most agro-industrial field crops. They have a large application in supplementary irrigation of cereals (wheat and barley) during the winter months.

FIGURE 9.1 - Spray boom on trolley in operation.

SYSTEM LAYOUT AND COMPONENTS

In this system a boom, equipped with low-pressure rotary spray nozzles, applies the irrigation water to the plants overhead in the form of fine sprays. The boom is a single travelling irrigation sprayer pipeline hanging above ground. It is mounted on a trolley and fed from a long flexible PE hose/pipe connected to a large reel (drum). The reel is placed on a swivel-base mounted on a wheeled cart (Figure 9.2). It is equipped with a controller and other optional devices for the proper functioning of the system. For operation the whole compact unit is transferred (towed), with aid of a small agricultural tractor, and connected to the hydrant or other pressurized source of water. The trolley with the spray boom folded (or unfolded) is towed to the length of the hose up to 400 m at the far end of the field area to be irrigated.

The system operates with the water pressure. The power is transmitted from the machine's turbine drive system on to the reel over a gearbox and a chain drive. During operation the water turbine driven reel pulls automatically the boom across a large strip area irrigated and rewinds the hose/pipe back on the drum. The retraction speed is regulated from 2 to 60 m/h. When the hose is completely retracted the reel stops winding automatically. The swivel base can be turned up to 180° and the boom and hose are towed to another direction of the field, where the irrigation operation is repeated to a new area of the same field. Then the whole compact unit can be towed to another field or stored.

FIGURE 9.2 - The boom, the PR pipe and the reel machine.

The Boom with the water emitters (sprayers) and the trolley

The boom pipeline consists of 2–3 inches (50–80 mm) hot dip zinc galvanized steel pipes in approximate lengths of 4 m, foldable, with reduced diameters at both ends. It is mounted on a two or four wheels cart/trolley for bigger booms, made of high quality steel and aluminium with swivelling turntable with supports and safety device for transporting on the traveller. The boom is hanging above ground at 1.3 m to 2.5 m hydraulically adjustable height achieved by means of special mechanism on the trolley. Burdening weights continuously balance the boom parallel to the ground. The boom is quickly and easily folded and unfolded by one operator and its full length (width) varies according to the model from 15 m to 50 m. End spray nozzles or impact-rotating sprinklers at both boom arms can increase the effective wetting width (irrigation strip width) significantly (Figure 9.3).

The sprayers are mostly the deflector type and with grooved plates stationary or rotating for longer effective water throw radius up to 6 m, as in the Center Pivots. The size of nozzles varies from 2 mm up to 5 mm and the operating pressure ranges from 0.7 Bars to 2.5 Bars with flow discharges from 100 l/h to 4.5 m^3/h. They are the full circle (360°) and half circle (180°) and they are placed along the boom line on the lower part facing the ground at frequent spacing of around 1.5–2 m apart for efficient overlapping. Near the trolley half- circle sprayers of deflector type with reduced throw radius are placed to direct the water in front of the boom, away from the wheels and to preventing the trolley from being dug into the wet soil, thus enable backwards move on dry soil.

FIGURE 9.3 - The full-circle and the half-circle Sprayers.

The Flexible polyethylene pipe

This is made from reinforced special virgin HDPE and it is high pressure PN 10 Bars minimum. It is specially developed for reel machines and its size varies according to the system from 75 to 110 mm in lengths of 300 to 500 m. One end is connecter to the reel machine and the other to the boom line. It is the irrigation water conveyance and the boom line feeding-pipeline.

The Reel machine

According to the model and the size there are two, three or four wheel, heavy structural frame undercarriages, with turntable base on which the pipe revolving reel (drum) can be swivelled mechanical or hydraulic through 270° and turned to any desired position. The whole machine stays fixed immovable even at high pulling forces during operation. The reel drum is about 2 m diameters. It is equipped with a full-flow turbine drive system and a four stage gearbox, a controller (electronic or mechanical) an over-pressure shut-off valve, a tachometer and a feeler bar for the exact regulation of retraction speed. The reel is safe during operation, start-up and shut-off. It stops automatically after the pipe pull-off (Figure 9.4). The winding mechanism guides the PE pipe precisely and evenly through all windings and layers without bends or bruises on the rolled inner drum. The system can be interrupted at any time of the operation with the use of a gearshift lever. When needed the PE pipe can be rolled up quickly. A blow out system empties the water left in the pipe with the completion of irrigation. At the end of the operation the boom is easily folded and the trolley is lifted hydraulically into the transport position and loaded on the reel carriage. The whole structure is made of hot dip galvanised steel.

FIGURE 9.4 - The reel machine with the PE pipe.

DESIGN CRITERIA AND CONSIDERATIONS

The travelling Spray Booms are movable systems and can easily be transferred from one field to another towed by a tractor. The most convenient size the farmers are using is the ones with flow discharge of 25–35 m^3/h at 3.0–4.5 Bars operating pressure (connecting pressure), 35–50 m wetted strip width and area covered per position 0.8–1.5 ha, 10–12 hours of operation with total precipitation 40–50 mm irrigation depth. One irrigation machine can easily cover 25 ha with supplementary irrigation during the winter months. This practice is usually applied in cereals during drought periods. Each model can be modified for variable flow and irrigation strip width by the replacement of the sprayers of the boom.

Area, and Topography

The area should be a plain agricultural field of regular shape. The system can be towed and moved to a next position nearby and so on (Figure 9.5). The Spray booms can operate on uneven ground, however, level lands are recommended and uniform sloping fields with slopes up to 1 percent. Undulating topography may produce a lot of difficulties especially where runoffs occur.

FIGURE 9.5 - Towing the compact machine.

Soil

The soil should be of medium texture with high infiltration rate >15 mm/hour good internal drainage and water holding capacity.

Water availability

The source of water can be a tube-well, a river, a small water tank. The system can be fed with water from hydrants placed at various points on the farm plot boundaries. The water pressure should be adequate for the system normal operation and in any case not less than 3.5 Bars. Otherwise a booster pump is needed, to delivering the flow at the required pressure connected directly to the system inlet. The system inlet will be connected to the hydrant or the pump outlet through a quick coupling flexible hose. For every Spray boom position a hydrant is needed. The water source should be as near as possible too the field.

Water quality

The water should be clean and free from suspended solids and other impurities, of normal pH 6.5 to 8.4, with no salinity hazard, sodium hazard and toxicity problems caused by bicarbonates, nitrates or boron. TDS should not exceed, if possible, 1 500 mg/l (ppm), SAR < 12, RSC < 1.25 meq/l, Boron content < 0.7 mg/l, Chlorides <200 mg/l, Nitrates (NO_3) < 100 mg/l and low content Bicarbonates (HCO_3).

Kind of crops

The field crops to be grown, among others, under Spray boom irrigation and their growing season are the same as with the Center pivots as follows:

- *Winter Crops:* Wheat, barley, mid-October/mid-November (sowing) to May/June (harvest), Chickpeas, March (sowing)–June (harvest), Lentils, May (sowing)–June (harvest).

- *Industrial Crops:* Soybeans, March/April (planting)–October/November (harvest). Maize, March/June/July (planting)–July/October (harvest), Sunflower, March (planting)–June (harvest).

- *Other Crops:* Leafy Vegetables, early spring; Potatoes (spring, autumn); Watermelons March/June-July/October (harvest); Groundnuts, April–Sept.; Lucerne, perennial.

SPECIAL CONSIDERATIONS AND IRRIGATION SCHEDULE

It must be underlined that this system is completely different from the conventional sprinkler irrigation systems as it is concerned with the intensity of irrigation water application. In the conventional stationary systems the rate of precipitation (P) is determined by the soil infiltration rate and depends on the sprinkler discharge (m^3/h) and the sprinkler spacing (m) along the line and between the lines. The depth (D) of water application (irrigation dosage) is determined by the irrigation schedule and depends on the hours of operation. E.g. when the precipitation rate (P) is 12 mm/h and the operation hours 3.5 then the depth of irrigation application (D) is 12 mm/h x 3.5 h = 42 mm.

In the travelling spray booms the rate of precipitation P is nearly identical to the depth (quantity) of water needed per unit of area per irrigation. This quantity is applied at once, like in the surface methods and not for a given a span of time (duration of application). In other words the area irrigated simultaneously is limited to the small area wetted at each moment by the boom sprayers, which along the travel deliver the whole amount of water needed. Then the intensity of precipitation is very high and ranges from the 50 percent to 100 percent the depth of irrigation application. It is obvious that run-off and water paddling is unavoidable in many fields of moderate and low infiltration rates, despite the manufacturers/suppliers opinion. The irrigation water dosage cannot be applied partially like with the Center Pivots by repeated travels, as it is not practical to the farmers.

The higher the retraction speed of travels the lower the depth of water application and vice/versa. Other important factors are the area covered per shift and the operating hours of the system. So, for the calculation of the irrigation program the following two formulas can be used, (1) for the determination of the retraction speed and (2) for the area irrigated per shift (setting):

$$S = \frac{Q \times 1000}{WD} \quad (1)$$

$$A = \frac{WSH}{10000} \quad (2)$$

where:

- **S** , retraction speed meters/h,
- **Q** , system flow m^3/h,
- **W** , irrigation strip width meters,
- **D** , depth of irrigation mm (irrigation dosage)
- **A** , area per setting ha and
- **H** , hours of system operation

For the irrigation program the users of the spray booms can use the above formulas, or make use of the suppliers tables prepared for this purpose (Table 9.1). These tables contain the calculated depth of water application at various retraction speeds of the boom, based on the system flow and the irrigated strip width.

TABLE 9.1 - Example performance table for boom retraction speed and precipitation

System flow m^3/h	Strip width meters	Depth of irrigation application per setting			
		20 mm	30 mm	40 mm	50 mm
		Retraction speed m/h			
17.5	50	17	12	9	7
30	50	30	20	15	12
37	50	37	25	18	15
45.9	50	46	31	23	18

The figures of the table are calculated by the use of the above given formula (1) for the retraction speed determination.

The system is recommended for deep-rooted crops with irrigation dosage of around 40 50 mm. The use of the right sprayers with fine drops at the right size and distribution is of major importance, as the very small drops are drifted by winds and distort uniform application pattern, whilst larger drops may have impact on the soil surface and influence the infiltration rate.

COST

The cost for a complete traveller spray boom system varies according to the size of the unit. The general characteristics for a relatively moderate size machine are, 25 40 m/h system water flow, 3.0 3.5 Bars connecting pressure, 75 mm flexible HDPE hose 300 m length, 50 meters effective wetted strip, 1.5 ha average area irrigated per setting. For this unit the cost is approximate US$12 000.

ADVANTAGES

- Automated complete irrigation system in a compact movable unit.
- Irrigation efficiency 80 percent
- Considerable savings in labour
- Fine precipitation improves soil structure
- Simple to handle

- No pipe networks installations within the field (Figure 9.6)
- Gives the farmer practical solutions to many irrigation problems
- Ideal for supplementary irrigation of large and remote fields (Figure 9.7)

FIGURE 9.6 - Spray Boom in operation.

DISADVANTAGES

- High initial purchase cost
- High intensity of precipitation resulting into runoff and water paddling
- Not recommended for heavy soils of fine texture with low permeability
- Transportation from one field to another is made by a tractor (Figure 9.8).

FIGURE 9.7 - Spray Boom machine.

EXAMPLE DESIGN – Spray Boom in alfalfa

Area and crop

The field is a rectangular area with dimensions 100 m along the road x 280 m, more or less plain and uniform slope, total area 2.8 ha. It is planted with alfalfa, perennial, deep-rooted forage crop tolerant to salinity.

Soil, water and climate

The soil is of medium texture, with high infiltration rate of > 15 mm/h and Sa 120 mm/m. The source of the irrigation water is a deep tube-well of 30 m3/h capacity at 3.2 Bars pressure output. The water is slightly saline with TDS 1 400 mg/l with no sodium hazard and any other problem. The climate is semi-arid with warm summer and highest evaporation 9 mm/day in July August.

Crop Irrigation requirements and irrigation program

Alfalfa is a perennial crop and grows for 3 to 4 years in climates with mild winters. Crop water requirements are high ranging from 900 to 1 500 mm per season. There is a variation in the water consumption (kc values) during the irrigation season that depends mainly on the timing of the cuttings and the dormancy period, which is practiced by the farmers during the high temperature months. With average kc 0.95 and 80 percent irrigation application efficiency the gross irrigation requirements in this example case amount to 1 350 mm. This total depth (amount) of water is applied in steady irrigation dosages of 50 mm, with a total number of 27 irrigations per season.

System characteristics layout and performance

The selected Spray boom machine is of low connecting pressure around 3 Bars and flow 30 m^3/h to match with the available source. The length of the boom is around 32 m with 50meters effective irrigation strip width. The flexible HDPE hose is 75 mm and 300 m long. The Spray boom is towed and placed at one end of the field 280 m apart and the Reel cart on the other, near the hydrant, with the PE reel pipe laid unwinds on ground. When operated the pipe is re-winded and the spray boom on the trolley is dragged backwards irrigating a 50 m strip. With two spray boom settings (50 m x 280 m) the whole area is covered.

The application of 50 mm depth per irrigation is achieved with the 30 m^3/h system flow at a travel speed of 12 meters/hour (see table or make use of formula). The time required per setting is 280 m ÷ 12 m/h = 23.3 hours. With 8 operating hours per day three days are needed per setting and six days to fulfil irrigation. In summer time the daily operating hours can increase to 10 or 12, so the time to fulfil irrigation can be two four days, i.e. two days per setting. According to the irrigation program the total number of 27 irrigations will be given, 2 in April, 3 in May, 5 in June, 5 in July, 5 in August, 4 in September and 3 in October. During the peak demand the machine will operate for 20 days per month to meet the requirements (Figure 9.9).

FIGURE 9.8 - Layout (setting) of Spray Boom machine.

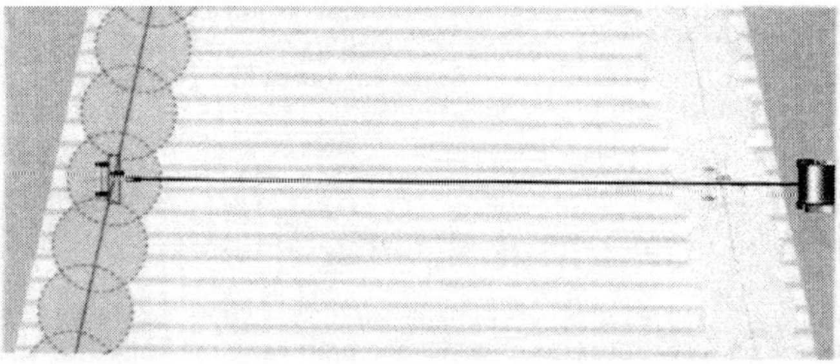

TECHNICAL SPECIFICATIONS (MINIMUM REQUIREMENTS)

Description and General Requirements

One Traveller Spray Boom Irrigation Reel type machine, integrated unit to be used as a single whole irrigation system as follows:

- Area cover per position: 1.2 to 1.8 ha.
- System flow and connecting pressure: 30 m/h at 3.0 to 3.5 Bars.
- Type of Systems: Travelling mechanised.
- System's Component parts: a) Reel machine on hydraulic travel wheeled cart, b) HDPE reel Pipe 75 mm c) Boom with sprayers mounted on trolley.
- System layout: The Spray boom is placed at one end of the field and the Reel cart on the other with the PE reel pipe laid unwinds on ground. When operated the pipe is re-winded and the spray boom on the trolley is dragged backwards.
- Total Length of reel pipe: 300 m.

Reel (assembly)

- Travel cart assembled from best quality steel and aluminium, with adjustable track width. Easily moved.
- Frame mounted on 2–4 wheelbase with large size tyres.
- Hydraulic or mechanical swivel aid for turning in any desired position on level and unlevelled ground.
- Machine supports with mechanical winch.
- Completely hot-dip galvanised travel cart and turntable.
- Revolving pipe Reel with integrated supports, swivel reel around 2 m
- Adjustable retraction speed with tachometer for digital speed indication.
- Equipped with drive mechanism powered by water pressure (full-flow turbine, pressure gauge, 4 speed gearbox, feeler bar, safety guards.)
- System for asymmetric and symmetric pipe guidance.
- Blow out system for residual water in the pipe.
- Hydraulic automatic lifting device for boom assembly and supporting legs
- Built-in automatic shut-off
- Arrangement for easy coupled to hydrant through connecting hoses 10 m.

Boom and Trolley

- Swivelling boom arms.
- Hot zinc plated or aluminium.
- Mounted on Trolley with wheels.
- Rotator Sprayers full-circle with fine drops and large radius coverage suitably placed at even frequent spaces; half-circle pattern near the trolley; end sprinklers.
- Effective irrigation strip width 50 to 55 m (Figure 9.10).
- Boom easily folded and unfolded.
- Adjustable boom height above ground 1.2 to 2.5 m.

FIGURE 9.9 - Variable irrigation strip widths with different sprayers arrangements.

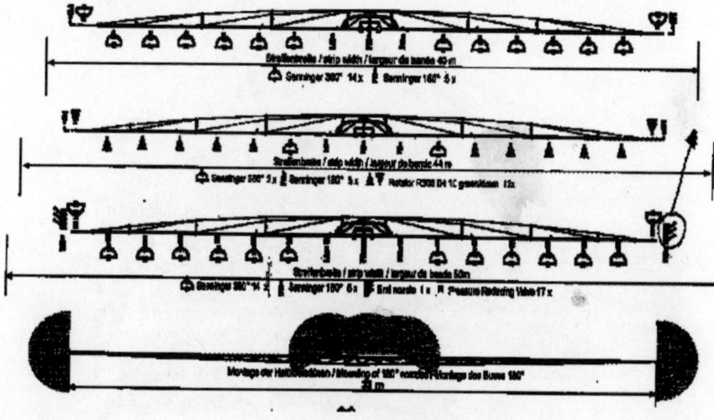

CHAPTER 10: The center pivot irrigation systems

INTRODUCTION

The center pivot system consists of one single sprayer or sprinkler pipeline of relatively large diameter, composed of high tensile galvanized light steel or aluminum pipes supported above ground by towers move on wheels, long spans, steel trusses and/or cables (Figure 10.1). One end of the line is connected to a pivot mechanism at the center of the command area; the entire line rotates about the pivot. The application rate of the water emitters varies from lower values near the pivot to higher ones towards the outer end by the use of small and large nozzles along the line accordingly.

The center pivot (CP) is a low/medium pressure fully mechanized automated irrigation system of permanent assemble. It has become very popular in the Near East region in recent years for irrigation of most of field crops, cereals, legumes, forage and vegetables. It is also used for supplementary irrigation for rain fed grain. The cost of each system unit is relatively high and is therefore best suited to large irrigated farms. The area covered can be from 3.5 ha to 60 ha, according to the size of the CP, and the larger the area the lower is the cost of the system per unit area.

FIGURE 10.1 - The center pivot.

SYSTEM LAYOUT AND COMPONENT PARTS

The typical center pivot system consists of a single long **irrigating pipeline** attached to a **central tower**and moves slowly over the field in a circular pattern and irrigates the plants with **sprayers, or sprinklers** placed on it at frequent spacing, The central tower with **a pivot mechanism and main control panel (electric) is** anchored to a small concrete base at a fixed water supply point (hydrant) at the center of the field. The entire irrigating pipeline is supported above ground by **"A" frame towers** move on wheels, **long spans**, steel trusses and/or cables; the end of the pipe is **overhung** with a sprinkler gun. The whole system rotates slowly, at a typical speed (last span) of 2–3 m/min., around the fixed pivot, self-propelled, applying water in the form of overhead spray irrigation and covers the area in a circular pattern. The **drive system**features small individual power units mounted on each wheeled tower. These units are electric drive, but can be hydraulic (water, oil) or mechanical drive. An automatic **alignment system** keeps always the irrigating pipeline straight (Figure 10.2).

The typical CP systems can be fixed permanent installations or movable/towable type with the central tower based on wheels or a skid, easily move from one field to another. The Linear Center Pivot is another common type towable system, which can irrigate rectangular or square shaped fields using a canal water resource parallel to the travel direction. Computer aided management systems (CAMS) models (MULTICENTER) and self-propelled are now available and the whole installation can be managed through a remote control for towing from one place to another. Corner systems are also available for irrigation of square, rectangular and odd shaped fields. Monospan systems are also available for small fields.

FIGURE 10.2 - A center pivot overview.

The Pipeline

The long irrigating pipeline (Lateral) with water emitters (sprinklers, bubblers, or sprayers) can be from 140 to 250 mm diameter, according to the system flow and the length; standard sizes of approximate 160 mm (6 inches) and 200 mm (8 inches) are very common. The length of the pipeline can be from 50 to 750 meters according to the design. It is made of high tensile material galvanized light steel, or aluminum, with extra strong couplers to stand the system operating pressures.

The pipeline is placed on wheeled "A frame" towers of typically 3 meters minimum height above ground and spaced 35–55 meters distance apart (length of spans). The common or "standard" length of spans is 40 m. Truss rod arches maintain the even distribution of weight and loads between the towers. On level ground the ground clearance varies from 2.75 to 4.5 meters for high profile machines. The spans are equipped with flexible joints at the ends allowing the pipeline to articulate and to allow side-to-side, up and down and rotational movement with no stress on the pipeline.

The water emitters

The water emitters, computerized sized and spaced for high uniformity of application, are mounted on the pipeline at spacing of 1.5 to 3.0 m, and 6 m approx. according to the type and coverage of the sprayer emitters, and operate when the system is in motion. The emitters in the past were full-circle rotating sprinklers. Since the early eighties the Low Energy Precision Application (LEPA) mode is using sprayers, bubblers or angle mist sprayers, fitted on flexible "hose drops" hanging down from the lateral at a height above ground of about 20 to 45 cm for the bubblers and 1.0–1.8 m for the sprayers. The "hose drops" are connected to the pipeline by a "gooseneck" or furrow arm and operate at lower pressures of about 0.5 to 1.5 bars. Goosenecks and drops are usually installed alternately on each side of the Lateral to even stresses on the line when used on high profile crops. There are several models of sprayers with excellent performance, long radius and uniform rain precipitation. Pressure/flow regulators are used in most cases. The discharge rate of the emitters along the pipeline is not the same along the line, but varies from lower values near the center to higher ones towards the outer end by the use of small and large nozzles along the line accordingly and sometimes variable spacing. Good overlapping is essential. Part circle sprayers are used near the towers to avoid over wetting the area along the wheels. The most common sprayers in use the last few years are the Senninger's wobblers and the Nelson's rotators.

A sprinkler gun is placed at the end of the overhang pipe (Figure 10.3) and may increase the length of the line by ¾ of its sprinkling radius, yet under low pressures rotating sprinklers have poor performance. Operating pressures along the line are low, that makes the system sensitive to pressure changes caused from friction losses or differences in elevation and uneven ground. Installation of pressure gauges for frequent monitoring is important to ensure proper flow and good application efficiency and uniformity. A pressure gauge is always needed at the last drop of the pipeline. Since 0.4 Bars pressure regulators often used with LEPA need at least 0.65 bars at the inlet to function properly, then maintaining 0.65 bars at the end of the line will ensure that all regulators are operating properly. The inlet pressure needed for the CP normal operation is slightly less than 3 bars.

BOX 10.1 - Low Energy Precision Application

Low Energy Precision Application (LEPA irrigation) is defined as: a low pressure irrigation method for uniformity applying small frequent irrigations at or near the ground level to individual furrows (usually alternate furrows) with a mechanical move system accompanied by soil-tillage methods or tillage plus crop residue management to increase surface water storage capacity. It is used on-farm with small diameter bubblers located about 0.3 m above ground and with socks or sleeves discharging water directly into furrows.

Spray irrigation is defined as: the application of water by a small spray or mist to the soil surface, where travel through the air becomes instrumental in the distribution of the water. It was developed to reduce the droplet evaporation and drift inherent with impact sprinklers. The sprinkler manufacturers have developed many types of spray heads and today many combinations of pressure regulators and spray heads, nozzles and deflector plates are commercially available.

Today the CP irrigation systems are mainly available in three types: a) The LEPA sock, b) The LEPA bubbler, c) The sprayer.

The bubbler mode produces an umbrella shaped pattern approximately 0.4 to 0.5 m in diameter, which minimizes wind effects and only wet part o the soil surface. This so called true LEPA system as developed and introduced by Lyle and Bordovsky (1981). The spray mode produces a horizontal spray approximately 2.5 to 3.3 m in diameter, which wets the entire soil surface. The so-called chemigate mode produces a 60° upward spray, which sprays the underside of the leaves). Lately new sprayers have been developed.

Commercial LEPA heads and nozzles, which can operate in the three different modes by changing the pad and hood positions, are now available. So, the same product can be used for pre-irrigation and germination, irrigation with the bubbler mode and chemigation.

FIGURE 10.3 - Hose drops with sprayers.

The Central Tower

This is a pyramidal structure of about 3.5–4.5 m height, built up with galvanized steel angular profiles and anchored on a concrete square platform. This structure has an access ladder. It is the head of the system and carries all equipment necessary for the control of the system, such as the system water fed up-going piece of pipe with the elbow on the top and inlets for fertilizer injection; the Collector ring, the Central Control Panel.

The CP System Control

A modular control panel, protected in a cabinet, is installed on the pivot central tower and enables handling of the irrigation machine and the programming of irrigation, thus the control of flow and pipeline movement–operation time and speed/time per lap (Figure 10.4). A voltmeter and several pilot lights indicate control tension, support tower alarm and luck of pressure. Automatic starter, position stop device, automatic shut-off and hour counter are included too in a standard control panel. Manufacturers offer several models of various level of control including full telemetry control with the use of cellular telephone.

FIGURE 10.4 - The Central tower and the control panel.

Driving

The CP irrigation system is fully mechanized automated. The drive system features small electric motors (standard HP 0.75) mounted one on each two-wheel tower. Whereas electric power is not available a generator is attached to the system. An automatic alignment system keeps always the lateral in a straight line. The distance that is covered by each tower varies, as the distance around the circle is greater at the far end of the pipeline and smaller near the center. The towers (Figure 10.5) do not move continuously, but in a series of starts and stops controlled by the movement frequency of the outer guide tower. The percent time set the fraction of time that the guide tower motor operates during each movement cycle, thereby setting the rate of revolution and the application rate. Micro-switches in the alignment mechanism operate the motors of interior towers to keep the system properly aligned.

FIGURE 10.5 - The Support towers.

The generator

In most cases and especially in areas without electricity facilities a small generator set attached to the central tower, diesel driven with fuel tank, provides the electric power for the system operation.

COST

The cost of each system unit is relatively high and varies from US$ 2 500–US$750 per ha depending on the size of the area. It is best suited to large irrigated farms. A relatively small to moderate unit with four spans pipeline of approximate 220 m radius for a 15 ha area per position costs around US$35 000.

ADVANTAGES

- High irrigation application efficiency 75–85 percent resulting into water savings, with the absolute control of the irrigation water from the source to the plants (Figure 10.6).
- Higher application uniformities as compared with other sprinkling or spraying systems because of continuously emitters.
- With the completion of one irrigation the system is at the starting point.
- Savings in labour and fuel
- Reduced tillage and associate costs
- Salinity control. Essential leaching of the root-zone at the end of the season is highly efficient with CP
- Supplementary irrigation in rain fed grains in drought periods.

FIGURE 10.6 - High and Low-profile structures.

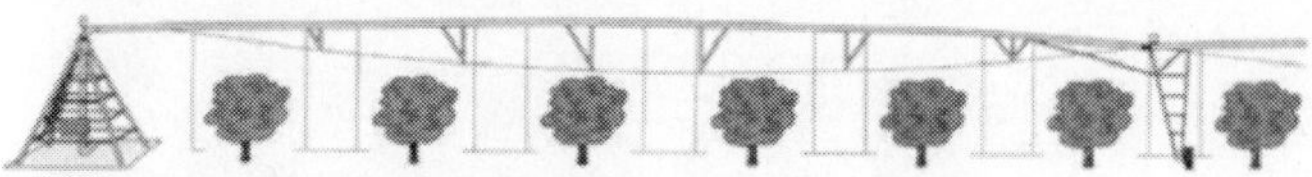

Ultra-High Profile and High-Profile Pivot and Drive Units
Provide increased crop clearance.

Low-Profile Pivot and Drive Units
When standard clearances are not required.

DISADVANTAGES

- High initial purchase cost.
- Not practical for small holdings

DESIGN CRITERIA AND CONSIDERATIONS

Design and erection

The manufacturers/suppliers are responsible for the design of the mechanical structures and the drive mechanisms for CP systems, since a specific knowledge of all the system components and features is required. They are also responsible for the erection of the machine (Figure 10.7). A turnkey system is provided to the end user. They also provide operator's manual that gives detailed steps necessary to prepare a CP system for operation at the beginning of the season, the maintenance during operation and the things that should be done to prepare the system for long periods when it is not used. New CP machines seldom need repair parts, but a source of repairs will be needed for several years after the installation. There are special requirements for transport (Figure 10.8), storing, loading/unloading the CP and erection of the systems.

FIGURE 10.7 - The erection of a center pivot.

FIGURE 10.8 - Towing the system.

Application rate and frequency of irrigation

All systems and irrigation methods are planned to apply water at rates equal or less the soil infiltration rate. CP is designed to be for the most part, independent of soil intake rate. Instead, CP design is based on the application volume per irrigation not exceeding the surface storage volume. With spray heads the wetted diameter greatly exceeds that with bubblers, however surface runoff occurs and reduce the application efficiency as well, as the uniformity of application, due to the redistribution of water in the field. The travel speed is important factor in designing these systems. In most cases the CP must run at a greater travel speed to compensate for the additional water reaching the ground. The irrigation frequencies are shortened and the irrigation application smaller. Irrigation Detailed irrigation programs should be prepared on the spot. In any case the in the semi-arid regions CP irrigation systems should be designed for crop requirements up to 8–9 mm/day at peak demand and increased daily operating hours during this period.

Special Considerations

The type and kind of soils – their topography, water holding capacity, and infiltration rate – are among the principal factors controlling the successful application of the CP. A major problem is that this type of CP system results into a very high precipitation rate, because the small emitters (sprayers and bubblers) have very small diameter coverage, (higher amount of water over a smaller area). This usually leads to the soil intake rate being exceeded, thus causing water run-off, or water ponding. In such cases additional runoff prevention practices are needed. CP systems operate best on soils with high infiltration rates that absorb the water at the point of impact and the nearby area. Runoff usually collect in low areas and log down the towers wheels. It tends to concentrate and create erosive streams. The one method to reduce runoff includes small dikes between the rows and the formation of small basins that keep the water where it falls, allowing it to be absorbed into the soil. Remedial tillage is necessary.

The CP systems operation are more trouble-free on level lands and on uniform sloping fields with slopes up to 3 percent. Slope differences can be up to 20 percent in the radial direction on rolling ground. In the circular direction the pipeline can move up and down slopes of 15 percent on plain fields or with shallow furrows. Where furrows are more than 0.15 m deep the ground slopes should never exceed 10 percent. Undulating topography may produce a lot of difficulties especially where runoffs occur. Substantial runoff or translocation occurs on deep slopes under CP irrigation without remedial tillage even if the soil is very sandy and water application is light. Because of the higher application rates than the infiltration rates significant runoff can occur on slopes greater than 3 percent. This problem can be remedy by performing some modifications the soil surface, or by the use of sprayers with larger diameter coverage, thus to apply water over a larger area.

On fields with uneven gradients variations in pressure can adversely affect the uniformity of application if pressure regulators are not used at each nozzle. With the use of circular CP, approximately 21 percent of a square field remains without irrigation. With the use of end guns this area is reduced to 15 percent.

In heavy loamy soils when wet soil does not support the weight of the wheels and deep ruts are cut, filling the ruts with sand can provide temporary solution to prevent stalling of the pipeline. If the soil becomes sticky as well, the pipeline may bog down if the land becomes too wet.

The speed of the CP is of major importance as it affects the rate of precipitation, which should not be identical to the irrigation dosage (water application). Then the irrigation dosage can be applied in several laps. As a rule of thump the minimum water application per irrigation cycle should be taken above 30 mm.

When water is pumped into the CP system, water fills the Lateral pipeline and the "hose drops". Weight of the water causes the pivot to squat especially toward mid-span. Length of the "hose drop" should be regulated to account for the change so that all emitter heads will be same height above ground when the system is running.

The Site Selection Criteria

The criteria to the collection and preparation of the above information are based on the kind and the type of the irrigation system and its techniques. A thorough study of the previous text specifies the site selection criteria for the installation of the CP irrigation systems as follows:

Area, size and shape

The area should be relatively level. This practice is usually applied in cereals for supplementary irrigation during drought periods.

Topography

The CP irrigation systems can operate on uneven ground; however, level lands are recommended and uniform sloping fields with slopes up to 3 percent. Undulating topography may produce a lot of difficulties especially where runoffs occur.

Type of soil

The soil should be of medium to light texture with high infiltration rate >15 mm/h and good internal drainage.

Water and pressure availability

The source of water can be a tube-well, a river, a small water tank. But the circular CP systems should be fed from a hydrant placed near the pivot. A booster pump can deliver the flow at the required pressure. The system inlet is connected to the hydrant through a quick coupling flexible hose.

Water quality

The water should be of normal pH, free from suspended solids, salinity hazard, sodium hazard and toxicity problems caused by bicarbonates, nitrates or boron. TDS should not exceed 1 500mg/l (ppm), SAR < 12, RSC < 1.25 meq/l, boron content< 0.7 mg/l.

Fuel requirements

The CP systems are equipped with generators for driving the towers and booster pumps, both diesel engine driven. The fuel tanks must be connected with additional bigger tanks placed nearby for long uninterrupted operation of the CP systems.

Kind of crops

Winter Crops (Wheat, barley, Chickpeas, Lentils) Industrial Crops (Soybeans, Maize, Sunflower), Other Crops (Leafy Vegetables, Groundnuts, Watermelons, Lucerne etc.) (Figure 10.9).

FIGURE 10.9 - Overview.

EXAMPLE DESIGN

The design of the CP irrigation project is relatively simple and in any case easier than the conventional drip systems. This example design is a most difficult case. A summer field crop is selected in a semiarid area with high evapotranspiration and scarce water resources. The main steps are as follows:

- Determination of the irrigated area, the length of the CP radius and the crop main characteristics (root depth, growing season, critical stages etc.).
- Calculation of the crop irrigation requirements at peak demand and soil available moisture (water holding capacity)
- Irrigation schedule (dosage, daily operating hours and interval)
- System main characteristics and technical specifications

Area, CP radius and crop

The area is around 15 ha, with minor slopes. The length of the CP pipeline is 215 meters with effective wetted length with the end gun 218.5 m (circle radius). The periphery of the circular area is around 1 350 m (Circle circumference = 2 π r) (Figure 10.10). The area is planted with groundnuts (Arachis hypogaea); it is a sensitive crop and any water stress results into yield reduction; growing season April–September, 130 days approximately; effective root depth 0.65 m.; kc values 0.45 initial stage, 0.75 vegetative growth, 1.0 at flowering and yield formation, 0.75 at the late stage. Mid-season stage is around 45 days with kc 1.0 and falls into July/August with the irrigation water demand at peak.

FIGURE 10.10 - System layout.

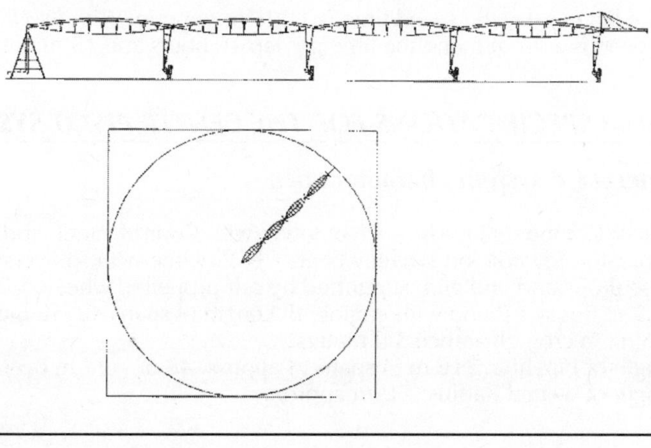

Crop water requirements and soil

Highest ETo in July-August is around 7.0 mm/day multiplied by crop kc 1.0, then ETc is 7.0 mm/day. The system application efficiency is 85 percent, so the gross irrigation needs are 7 mm ÷ 0.85 = 8.2 mm/day or 8.2 mm *x* 10 *x* 15 ha =1 235 m^3/day at peak demand period. Soil is light – medium texture with available moisture (Sa) 120 mm/m soil depth and infiltration rate > 15 mm/h. Based on the above data a detailed irrigation program (schedule) can be prepared.

Systems characteristics

The systems performances should meet the peak demand period as well as with the other growing period and the prevailing conditions with regard to the water availability, daily operating hours cost of fuel etc. The factors and data needed for the calculation of the system flow are the irrigation dosage and the daily water requirements. The dosage (d) = Soil available moisture mm/m (Sa) *x* moisture depletion percent (p) *x* root depth m (D). The frequency of irrigation in days is d mm ÷ Etc/day. The interval between irrigations (end of one irrigation until the beginning of the next) is the frequency in days minus the days required for fulfilment of one irrigation. The latter depends on the size of flow and the daily operating hours. In most cases the available flow dictates the basic system characteristics and performances. Irrigationists should take into consideration that the designed flow in practice is available only in 7 percent to 9 percent of the cases.

A system flow around 15–17 l/s (60 m^3/h) is selected and the minimum daily operating hours to meet the demand at the crop mid-season stage is:

$$1\ 235\ m^3/day \div 60\ m^3/h = 20.5\ h/day.$$

For one lap per day the speed time is 1.125 m per min. With two laps per day the speed is 2.25 m/h and the time per lap 10 hours and 15 minutes.

MINIMUM SPECIFICATIONS FOR THE CENTER PIVOT SYSTEM

The general CP system characteristics

- System Component parts: a) Pivot tower with Control Panel, and Generator, Towable on swivel wheels – b) Pipeline with sprayers on Hose drops and end gun, supported by self-propelled wheeled Towers (4), – c) Booster Pump with engine. d) Length of spans 40–50 meters. Minimum crop clearance 3.0 meters.
- Length of Pipeline: 216 m (4 spans of approx. 48 m + 24 m overhang)
- Length of wetted Radius: 219m approx.

- System Flow: 60 m³/h
- Pressure at the Pivot Inlet: 2.3 Bars approx. (Figure 10.11)
- Area coverage: 15 ha approx.
- Min. speed, Time per Lap, Precipitation: 0.3 m/min., 74 hours, 29.6 mm
- Max. speed, Time per Lap, Precipitation: 3 m/min., 7.4 hours, 2.96 mm

FIGURE 10.11 - Connection to the water source.

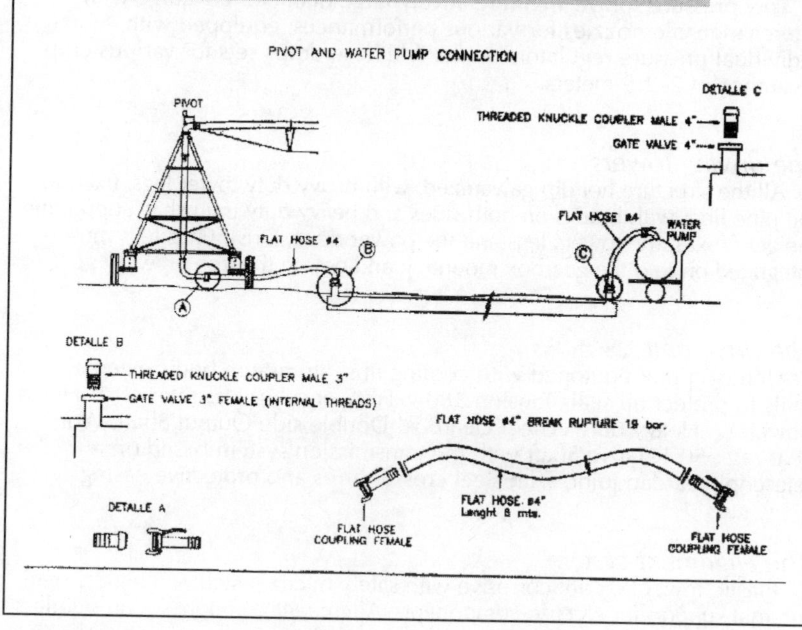

The technical specifications

The central tower

Heavy-duty structure, hot dip galvanized including the elbow. Braces on all sides and Access ladder for checking the Collector ring. Appropriate size of riser inlet tube for minimum friction losses. Tower on swivel wheels and heavy-duty feet in cases of anchoring the pyramid to the concrete base. Galvanized Main Control Panel support angles. Pressure gauge. Riser-pipe support angles. No vibration of the inlet water with high pressure.

The pipeline

Hot dip-galvanized steel zinc coated inside and outside. Pipe diameter approx. 141–160 mm for minimum friction losses, min. PN 10 Bars. Rubber gaskets, galvanized steel holders, short reinforced cardan type union of spans for great resistance, rubber union sleeves with steel clamps or aluminum couplings, flanged type of jointing of spans. All structure hot dip galvanized for preventing corrosion, unions with security bolts and nuts.

The water emitters

Low pressure sprayers with relatively large diameter coverage with interchangeable nozzles for various performances; equipped with individual pressure regulators on flexible hose drops sets for various crop clearance of 2–2.5 meters.

The driving towers

All the structure hot dip galvanized, with heavy duty tower legs, fixed to the pipe line, with braces on both sides and heavy duty train shaft supporting the gearboxes, the towers legs and the power drive. Drive units legs must be integrated on into the gearbox mounting and not on the free pipe.

The drive unit

Compact unit equipped with cooling fins, aluminum body, external dust seals to protect oil seals Tension 380 v. Frequency 50 Hz, rate: 40:1. Horse power ¾. Heavy duty Wheel Gearbox. Double side Output Shaft. Worm gear rate: 50:1. Drive Shaft with the transmission system based on a telescopic cardan joint, with steel crosses arms and protective casing.

The alignment system

Plastic Tower box anticorrosive with safety micro switches. Stainless steel internal support for electric components. Alignment components on stainless steel and aluminum. Individual electricity switch on every control box.

The control panel

Closed Box reinforced, double door, airtight totally isolated interior, Tension 380 V. Frequency 50 Hz. Including voltmeter, several pilots indicating control tension, support tower alarms and lack of pressure, control indicator forward and reverse running, forward and reverse selector, with or without water selector, automatic shut down (when pressure is insufficient), tension tester, hour counter, speed control, general switch. Possibility of stopping the electric generator electro valve or the water pump station in case of any problem during the pivot work. Outlet available for 24 and 110 V.

Collector ring

Totally sealed placed on top of the central pivot tower.

Wheels

Wheels for Irrigation use. First use High flotation wheels. Hot dip galvanized rims, with valve protection.

CHAPTER 11: Microsprinklers

INTRODUCTION

Microsprinklers are low capacity water emitters, sprinkler in type, but smaller in size than the conventional sprinklers and with flow rates up to 250 litres/h. They are placed on a relatively close rectangular or triangular spacing for the maximum overlap to irrigate potatoes, carrots, leafy vegetables, groundnuts and other densely planted field crops. This method is reliable, highly efficient, and easy to apply, operate and handle.

The system is a seasonal, low pressure, micro-irrigation solid installation which can be easily placed in the field and quickly removed (collected) at the end of the season.

SYSTEM LAYOUT AND COMPONENTS

The layout of the system consists of a head control equipped only with the regulating valves (shut-off, non-return, air) and a filter of about 40–60 mesh (200–300 microns). No injectors are needed as fertigation through this system is not a common practice among farmers.

The arrangement of the main and submain lines, hydrants and manifolds is the same as in other micro-irrigation piping networks.

The size of the manifold feeder lines should be 50–63 mm and in no case exceed 75 mm. Pipelines of 50–63 mm are recommended for flows of approximately 12–18 m^3/h when the water is distributed en route continuously.

The pipes used for the system's distribution network are mainly in rigid PVC (buried) or black HDPE (normally laid on the surface). Other kind of pipes are used also, such as layflat hoses and quick coupling galvanized light steel pipes.

The laterals are soft PE pipes, 20, 25 or 32 mm in diameter, according to the length, PN 4.0 bars, laid permanently on the surface. The microsprinklers are placed along the laterals at a spacing of 5–7 m, fixed 70–80 cm above the surface on iron rods inserted into the ground. They are connected to the laterals through a small flexible PVC tube 7–9 mm in diameter and 1 m long.

MICROSPRINKLER EMITTERS

These emitters are low capacity rotary sprinklers designed for low discharges uniformly distributed over the irrigated area in a rainfall pattern (Figure 11.1). Made of durable plastic, they have various operating mechanisms, and are usually compact without external moving parts. They have one low trajectory (jet angle above nozzle), quick rotating, 1.5–2.0 mm nozzle. The main performance characteristics are:

- operating pressure: 2.0 bars;
- flow rate (discharge): 130–250 litres/h (recommended 160–180 litres/h);
- wetting diameter (coverage): 12 m average;
- precipitation rate: 4–7 mm/h (recommended);
- filtration requirements: 40–60 mesh (300–250 microns) approximately.

A complete set consists of: a) the sprinkler emitter compact head; b) a 6 mm iron rod 1 m long; and c) a 7–9 mm flexible PVC tube with a barbed plunger for connection with the PE lateral line.

FIGURE 11.1 - Microsprinkler emitters in potatoes.

IRRIGATION SCHEDULING PROGRAMME

This system enables a high degree of control of both when to apply and how much irrigation water to apply. The restrictions imposed by the system are limited. Thus, there are more timing options in the irrigation scheduling programme. The vegetables are mostly shallow rooted crops, so the selected option is generally that of fixed depletion irrigation.

A gross application depth of 20–30 mm is common for potatoes and vegetables. The gross water requirements of a vegetable or a potato plantation vary from 300 to 400 mm in terms of depth. Thus, the total number of irrigations required is about 12–15, at intervals based on the cumulative evaporation.

DESIGN CRITERIA AND CONSIDERATIONS

In addition to the standard design criteria, such as area, crop, water supply, soil and climate, it is important to examine the system's special features and characteristics as these parameters influence the final decision.

Microsprinklers deliver the water in low application rates and in fine drops. Such drops easily drift in the air, even under low to moderate wind conditions. In order to ensure a high uniformity of application, the sprinkler spacing should be decreased and not exceed 50 percent of the diameter coverage, i.e. the sprinkler spacing along the laterals and between the laterals should range from 5 to 7 m. Hence, common spacings are 5 x 5 m, 5x 6 m, 5 x 7 m and 6 x 7 m. Furthermore, to mitigate the adverse effects of the wind, a relatively large number of sprinklers per unit area should be operated simultaneously. The operation shifts should be arranged so that the area irrigated at the same time is as compact as possible.

The system laterals are made of LDPE. Experience has shown that the optimum size of these pipes for this system is 32 mm as such pipes are easy to handle on site, to place, to remove, etc. Larger diameters are not recommended.

The maximum permissible length of various size laterals on level ground depending on the number of the sprinklers, the spacing and the flow rate is as follows (Table 11.1):

TABLE 11.1 - Maximum permissible length of laterals

Lateral size and spacing		160 litres/h		180 litres/h	
		No. of sprinklers	Lateral length m	No. of sprinklers	Lateral length m
20 mm	5 m	8	40	7	35
20 mm	6 m	7	42	6	36
20 mm	7 m	7	49	6	42
20 mm	8 m	7	56	6	48
25 mm	5 m	12	60	11	55
25 mm	6 m	11	66	10	60
25 mm	7 m	10	70	10	70
25 mm	8 m	10	80	9	72
32 mm	5 m	21	105	18	90
32 mm	6 m	20	120	17	102
32 mm	7 m	18	126	16	102
32 mm	8 m	18	144	15	120

All pipes LDPE 4.0 bars to DIN 8072 (inside diameter 16.0, 20.2 and 27.2 mm respectively).

COST

The cost for a complete installation of this system is approximately US$3 300/ha. The head control unit accounts for 8–10 percent of the total cost of the system; the plastic pipes (tubes), exactly 50 percent; and the low capacity sprinklers, nearly 35 percent.

ADVANTAGES

- Low labour O&M requirements.
- Flexibility and adaptability: the technology is simple and easy to adopt and manage. The safe transition from traditional surface methods to advanced micro-irrigation can be successfully accomplished through the installation of this kind of system.
- High irrigation application efficiency.

DISADVANTAGES

- High initial purchase cost.

EXAMPLE DESIGN – Microsprinklers with potatoes

Area and crop

The plot dimensions are 120 85 m (1.0 ha) planted with spring potatoes (Figure 11.2).

Soil, water and climate

Fine texture soil of good structure with a permeability of approximately 11 mm/h, and a high water holding capacity (200 mm/m). The water is supplied from a properly treated reservoir; it is clean, but slightly saline. The spring potato season is from January to May. In April the average pan readings are 3.3 mm/d; multiplied by the pan correction factor (0.66), this gives ETo = 2.18 mm/d.

Crop water requirements and irrigation schedule

Peak demand is in April and the kc value is 0.9. Then, ETc = 2.18 x 0.9 = 1.96 mm/d (net water requirements at peak). The system's application efficiency is 75 percent. In addition, an extra amount of about 15 percent should be applied for leaching salts. The peak gross irrigation requirements are:

FIGURE 11.2 - Low capacity sprinkler irrigation in potatoes (solid system). (Sample design)

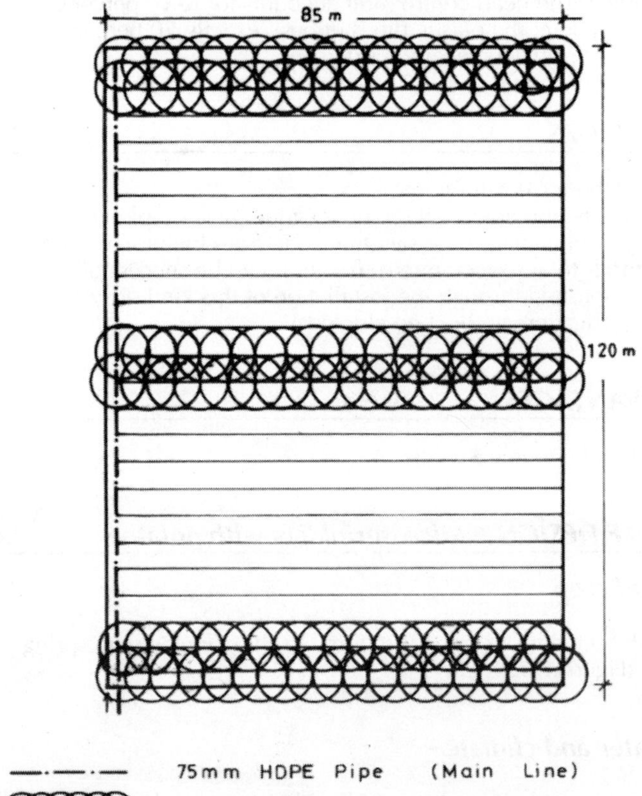

$$1.96 \text{ mm/d} \times 100 \div 75 = 2.61 \text{ mm/d} \times 100 \div 85 = 3.1 \text{ mm/d} \times 10 \times 1.0 \text{ ha} = 31 \text{ m}^3\text{/day}$$

The soil available moisture is 200 mm/m depth, the effective root depth is 0.35 m, and the maximum recommended moisture depletion is 40 percent. The maximum irrigation interval in April is:

$$200 \times 0.35 \times 0.4 \div 1.96 = 14 \text{ days}$$

The irrigation scheduling programme is arranged at a fixed depletion of about 20 mm (cumulative evaporation). Thus, the interval in April is: 20 ÷ 1.96 = 10 days. The gross irrigation application is 31.4 mm, which gives a gross amount of 314 m^3/ha per irrigation.

System layout

The system is a solid installation with all the sprinkler laterals permanently laid on the field. The head control unit is equipped with regulating valves and a screen filter of 60 mesh. There is only the main line, 75 mm HDPE, 6.0 bars, laid along the side of the field serving also as a manifold, feeding the laterals. The sprinkler laterals are 32 mm LDPE, 4.0 bars, connected with the mains through 2 ½ inches hydrants.

Sprinklers

160 litres/h at 2.0 bars, full circle, wetting diameter (coverage) 11 m. Spacing: 5 m along the lateral x 5 m between the laterals. Precipitation rate: 6.4 mm/h. Number of sprinklers per lateral: 17.

Lateral line discharge

2 720 litres/h. Total number of laterals: 24. Number of laterals operating simultaneously (per shift): 6. System discharge: 16.3 m^3/h. Number of shifts per irrigation: 4. Duration of application per shift: 4.9 h (4 h 50 min). Time required for one irrigation: 19.5 h (all shifts) (Tables 11.2 and 11.3 and Figure 11.3).

TABLE 11.2 - System's operating pressure

	bars
Pressure required at the sprinkler head	2.00
Friction losses in the lateral	0.20
Friction losses in the main line	0.35
Friction losses in the head control	0.50
Minor local losses	0.20
Total dynamic head	3.05

FIGURE 11.3 - Low capacity (micro) sprinkler system.

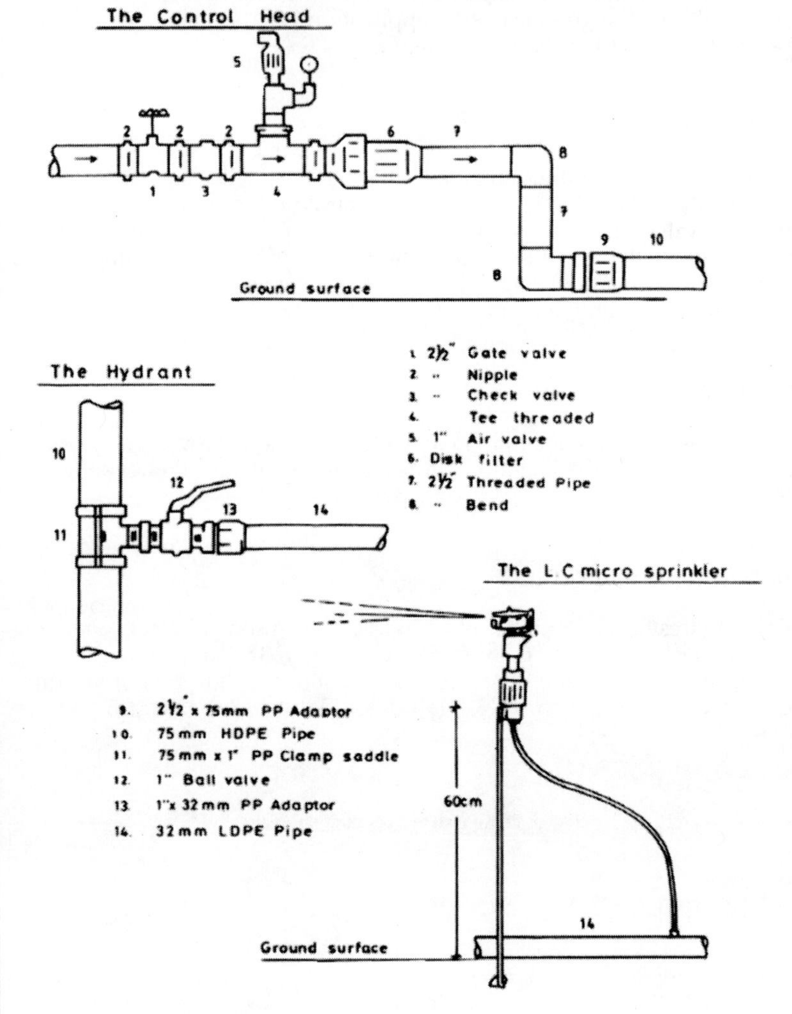

TABLE 11.3 - List of equipment needed for the installation

Item	Description	Quantity	Unit price US$	Total price US$
	System distribution network			
1.	75 mm black HDPE, 6.0 bars	120 m	2.60	312.00
2.	32 mm black LDPE, 4.0 bars	2040 m	0.65	1326.00
3.	75 mm x 2 ½ in PP adaptor	1 pc	9.00	9.00
4.	32 mm x 1 in PP adaptor	24 pcs	1.25	30.00
5.	75 mm PP end plug	1 pc	9.00	9.00
6.	32 mm PP end plug	24 pcs	1.25	30.00
7.	75 mm x 1 in PP clamp saddle	24 pcs	1.80	43.20
8.	1 in brass shut-off valve	24 pcs	3.50	84.00
9.	1 in nipple	24 pcs	0.40	9.60
10.	Low capacity sprinklers, full circle, 160 litres/h at 2.0 bars, 11 m wetted diameter, complete with supporting stake and connector tube	408 pcs	2.80	1142.40
	Sub-total			**2995.20**
	Head control			
11.	2 ½ in brass check valve	1 pc	15.00	15.00
12.	2 ½ in brass shut-off valve	2 pcs	13.00	26.00
13.	2 ½ in tee (galvanized iron or PVC)	3 pcs	3.50	10.50
14.	2 ½ in nipple	4 pcs	1.00	4.00
15.	1 in air valve (single automatic)	1 pc	12.00	12.00
16.	2 ½ in filter screen type 60 mesh	1 pc	180.00	180.00
	Sub-total			**247.00**
	TOTAL COST			**3242.20**

CHAPTER 12: Minisprinklers

INTRODUCTION

With this method a single minisprinkler emitter is placed for each tree. The emitter sprays water in a circular pattern under the foliage at low ratesover a limited area around the tree. This approach combines the principles and advantages of both sprinkler and localized drip irrigation.

Minisprinkler irrigation is a localized micro-irrigation approach which uses a low pressure system in a solid permanent or seasonal installation.

SYSTEM LAYOUT AND COMPONENTS

The control station can be as simple as possible. However, a filtering device is necessary as in all micro-irrigation installations. The fertilizer injector is not always needed, as many farmers prefer to apply fertilizer manually. Nonetheless, the arrangement of the equipment should be one which will enable the installation of a fertilizer injector at a later stage.

The mains and the submains can be of any kind of permanently assembled pipes, either on the surface or buried, with hydrants (2–3 inches) rising on the surface, or protected in valve boxes where buried.

The manifold (feeder) pipelines can be either surface-laid HDPE or buried rigid PVC pipes. Other kinds of pipes can also be used, such as layflat hose or quick coupling light steel pipes.

The laterals with the minisprinklers are laid along the rows of the trees near the trunks, one line at each row, with one minisprinkler per tree.

The lateral pipelines are generally (soft) 16, 20, 25 and 32 mm LDPE pipes, 4.0 bars PN. Buried small-diameter PVC pipes can also be used for the laterals, with longer connecting small plastic tubes, rising on the surface.

THE MINISPRINKLER EMITTER

The microsprayers used in tree orchards (Figure 12.1), called minisprinklers, spitters or micro-jets, are small plastic emitters of the static sprinkler type with a low-angle small water discharge in the form of fine drops which is uniformly distributed around the trees in a full or part circle pattern.

They can be of various mechanisms (capped with a rotating needle, non-capped with a swivel spreader, or with a deflector) with a wide range of flow rates and water diameters. All kinds have rather small flow sectional areas (nozzle diameter 1–1.7 mm approximately). The irrigation water needs to be filtered before it enters the system.

The main performance characteristics of the minisprinklers are:

- operating pressure: 1.5–2.0 bars;
- flow rate: 35–250 litres/h (generally 150 litres/h);
- wetting diameter: 3–6 m;
- precipitation rate: 2–20 mm/h (generally 4–8 mm/h);
- filtration requirements: 60–80 mesh (250–200 microns).

The minisprinkler heads are fixed to small plastic wedges or metallic rods 20–30 cm above the ground and they are connected to the PE laterals with 7–9 mm flexible plastic tubes 60–120 cm long and a barbed plunger. Thus, a complete minisprinkler emitter consists of the head, support wedge and connecting tube with plunger. All component parts are press-fit, interchangeable and easily assembled and dismantled.

FIGURE 12.1 - Irrigation of citrus orchard with minisprinklers.

IRRIGATION SCHEDULING

As in all localized micro-irrigation methods, the amount of water stored in the root zone is restricted as a result of the limited wetted soil volume. However, with this method the wetted volume of soil exceeds 65 percent of the total volume, there thus being no urgent need for very frequent irrigations unless the soil water holding capacity is very low.

The common practice is to irrigate at a fixed interval on a weekly basis and to apply the accumulated water requirements in the preceding days. With young trees, the irrigation interval is shorter, twice a week. Farmers in most arid and semi-arid zones apply water to their tree crops (citrus, guavas, avocado, etc.) as per Table 12.1.

TABLE 12.1 - Minisprinkler irrigation scheduling

Age of trees (years)	Litres per tree per day	Irrigation interval (days)	Average irrigation dosage per tree (litres)	Average operation time (hours)
1-2	8-15	4-6	60	0.5
3-4	20-40	6-7	200	1.6
4-6	50-60	7	380	3.0
7 plus	80-120	7-10	900	7.5

DESIGN CRITERIA AND CONSIDERATIONS

Minisprinklers are mainly used with intensively irrigated fruit trees (Figure 12.2). They can also be used with rainfed trees for supplementary irrigation. One emitter per tree is sufficient; therefore, the emitter spacings are identical to the tree spacings. The distance between the minisprinklers and the tree trunks is 30–50 cm depending on the age and size of the tree. For young trees, the minisprinkler heads can be mounted upside down to reduce the wetting diameter.

The emitter flow rate should be one which matches the existing conditions of water availability; the area; the number, age and size of the trees; and the number of irrigation shifts (irrigation programme). This is not a difficult task, considering the large range of minisprinkler flow rates available.

The minisprinkler emitters are short-path nozzle-orifice emitters with fully turbulent flow. Therefore, the variation in minisprinkler discharge is half the variation in the operating pressure; e.g. a 20 percent difference in pressure results in a 10 percent difference in discharge, which is considered the maximum permissible. Table 12.2 is based on this principle.

TABLE 12.2 - Maximum number of minisprinklers on the lateral and length of lateral on level ground. Minisprinkler flow rate at 2.0 bars

Lateral size and spacing (metres)		70 litres/h Max. no. of minisprinklers	70 litres/h Lateral length (metres)	120 litres/h Max. no. of minisprinklers	120 litres/h Lateral length (metres)	150 litres/h Max. no. of minisprinklers	150 litres/h Lateral length (metres)
16 mm	3	10	30	7	21	6	18
	4	9	36	6	24	6	24
	5	8	40	6	30	5	25
	6	8	42	5	30	5	30
	8	7	56	5	40	4	32
20 mm	3	16	48	11	33	9	27
	4	15	60	10	40	9	36
	5	14	70	9	45	8	40
	6	13	78	9	54	8	48
	8	11	88	8	64	7	56
25 mm	3	25	75	18	54	15	45
	4	22	88	16	64	14	56
	5	20	100	15	75	13	65
	6	19	114	14	84	12	72
	7	18	144	12	96	11	88

Note: All pipes are LDPE, 4.0 bars, to DIN 8072.

FIGURE 12.2 - Minisprinklers in trees.

COST

The total cost for the installation of the system on 1.0 ha, as in the example design, is US$1 634. A cost analysis shows that the head control unit costs US$470, i.e. 26 percent of the total. The same unit can serve an area of at least 3.0 ha. The most important cost item is the PE pipes (tubes), i.e. the system network, at US$864. This is 55 percent of the total cost for a 1 ha system. For a 3 ha complete installation, the pipes represent about 65 percent of the total cost.

ADVANTAGES

- High irrigation application efficiency. The amount of water is precisely controlled and only a partial area is wetted. No losses occur due to evaporation, deep percolation or runoff.
- Salinity control. The water movement through the soil profile is vertical downwards and the accumulated salts in the root zone can easily be leached to deeper layers.
- Flexibility and adaptability. It is the most flexible micro-irrigation system and is easily adopted and managed by farmers. The technology is simple and the range of equipment relatively low.
- Low labour requirements.

DISADVANTAGES

- High initial purchase cost.

EXAMPLE DESIGN – Minisprinkler with citrus lemon trees

Area and crop

The plot is 85 x 120 m, i.e. about 1 ha, planted with mature (12-year-old) citrus lemon trees in rows at a spacing of 6 x 6 m (Figure 12.3). There are 20 rows with 14 trees in each row for a total of 280 trees. The slope of the plot is 0.5 percent from west to east and from north to south.

Soil, water and climate

The soil is of medium texture with a permeability of approximately 4 mm/h, a water holding capacity of 22 percent, and soil available moisture of 150 mm/m depth. There are no salinity or toxicity hazards. The source of water is an existing tube-well with a safe output of 5 litres/s (18 m^3/h).

The water is of good quality with the electrical conductivity of the irrigation water ECw = 1.5 dS/m total salinity. The evaporation pan average readings in mid-July are 7.0 mm/d. The irrigation season is from April to October.

Crop water requirements and irrigation schedule

A pan reading of 7.0 mm/d multiplied by 0.66 (pan correction factor) gives an ETo of 4.65 mm/d. The crop factor kc is 0.65. Thus, ETc = 4.65 *x* 0.65 = 3.0 mm/d. The area shaded by the tree canopy is 70 percent and for calculation purposes it is taken as 82 percent. Therefore, the daily water requirements are: 3.0 *x* 0.82 = 2.48 mm/d net. With a system application efficiency of 80 percent, the gross daily irrigation requirements are: 2.48 *x* 100 ÷ 80 = 3.1 mm (31 m^3). If irrigation takes place every 10 days, the gross irrigation dosage is: 10 *x* 31 = 310 m^3.

FIGURE 12.3 - Mini sprinkler irrigation in fruit trees. (Sample design)

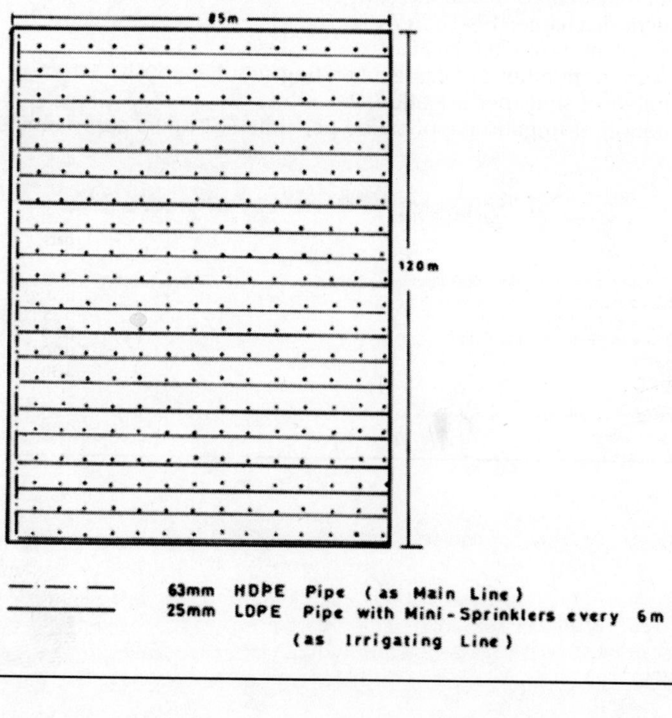

The maximum permissible irrigation interval in July with a 50 percent moisture depletion for a tree root depth of 0.6 m is: 150 x 0.6 x 0.5 ÷ 3.0 = 15 days. The irrigation frequency depends on many factors and in no case should exceed the maximum permissible irrigation interval.

However, as mentioned, the common practice is to irrigate every seven days. The seven-day interval accumulates gross irrigation requirements of 217 m^3, i.e. the irrigation dosage at peak demand in July.

SYSTEM LAYOUT, PERFORMANCE AND HYDRAULICS:

- One lateral irrigating line of 25 mm LDPE pipe, 81 m long, is laid along each row of trees, with one minisprinkler per tree, i.e. 14 minisprinklers per line. The laterals lines are connected directly to the main line, a 63 mm HDPE pipe laid along the plot boundaries, which also serves as a manifold (Figure 12.4).
- Minisprinkler performance: 120 litres/h flow rate full circle, 2.0 bars, 60 mesh, filtration required (Tables 12.3 and 12.4).
- Number of minisprinklers per lateral line: 14
- Lateral discharge: 1 680 litres/h
- System discharge: 17–18 m^3/h
- Total number of laterals: 20
- Maximum number of laterals operating simultaneously: 10
- Number of shifts per irrigation: 2
- Duration of irrigation application per shift: 6.2 hours approximately.

TABLE 12.3 - System's operating pressure

	bars
Pressure required at minisprinkler operation	2.00
Friction losses along the lateral line	0.35
Friction losses along the main line	0.25
Friction losses in the head control unit	0.50
Minor local losses	0.20
Sub-total	**3.30**
Difference in elevation	- 0.15
Total dynamic head	3.15

FIGURE 12.4 - Minisprinkler system jointing techniques.

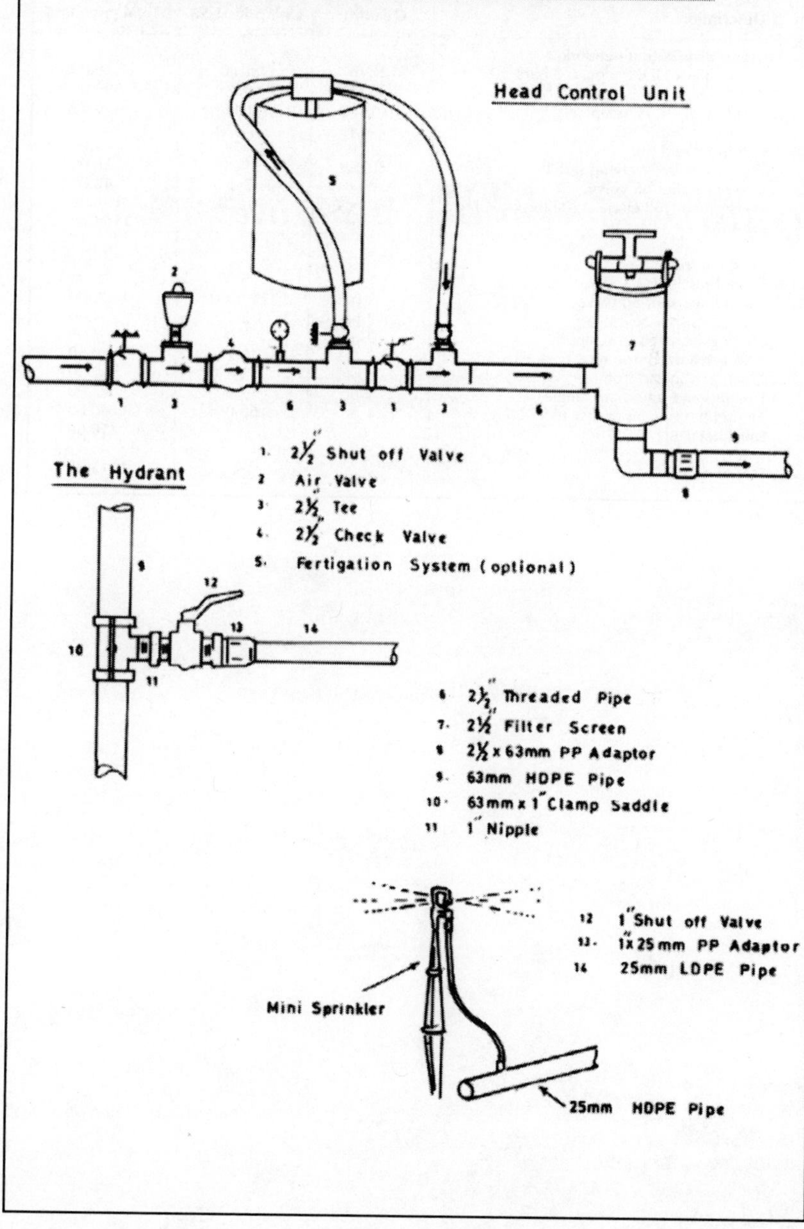

TABLE 12.4 - Equipment required for the system installation

Item	Description	Quantity	Unit price US$	Total price US$
	System distribution network			
1.	63 mm black HDPE pipe, 6.0 bars	120 m	1.80	216.00
2.	25 mm black LDPE pipe, 4.0 bars	1620 m	0.40	648.00
3.	63 mm x 2 ½ in PP adaptor	1 pc	6.00	6.00
4.	25 mm x ¾ in PP adaptor	20 pcs	1.00	20.00
5.	63 mm PP end plug	1 pc	6.00	6.00
6.	63 mm x ¾ in PP clamp saddle	20 pcs	1.30	26.00
7.	¾ in brass shut-off valve	20 pcs	2.30	46.00
8.	Minisprinkler full circle, 120 litres/h, 2.0 bars	280 pcs	0.70	196.00
	Sub-total			**1164.00**
	Head control			
9.	2 ½ in brass check valve	1 pc	15.00	15.00
10.	2 ½ in brass shut-off valve	2 pcs	13.00	26.00
11.	¾ in brass shut-off valve	2 pcs	2.30	4.60
12.	2 ½ in galvanized iron nipple	2 pcs	1.00	2.00
13.	¾ in galvanized iron nipple	2 pcs	0.25	0.50
14.	2 ½ in galvanized iron tee	3 pcs	3.50	10.50
15.	1 in air valve automatic single	1 pc	12.00	12.00
16.	2 ½ in filter (screen or disk type) 60 mesh	1 pc	180.00	180.00
	Sub-total			**250.60**
	TOTAL COST			**1415.00**

CHAPTER 13: Bubbler irrigation of trees

INTRODUCTION

Bubbler irrigation is a localized, low pressure, solid permanent installation system used in tree groves. Each tree has a round or square basin which is flooded with water during irrigation. The water infiltrates into the soil and wets the root zone. The water is applied through bubblers. These are small emitters placed in the basins which discharge water at flow rates of 100–250 litres/h. Each basin can have one or two bubblers as required.

SYSTEM LAYOUT AND COMPONENTS

The system layout is the typical one of all pressurized systems. It consists of a simple head control unit without filters and fertilizer apparatus. The mains and the submains are usually buried rigid PVC pipes, with hydrants rising on surface. The manifolds and laterals are also often buried rigid PVC pipes. The bubblers are placed above ground, supported on a stake, and connected to the laterals with a small flexible tube rising on the surface, or they can be fitted on small PVC risers connected to the buried laterals.

The difference between bubbler systems and other micro-irrigation installations is that whereas in the other installations the lateral lines are small (12–32 mm), the bubblers are usually 50 mm (due to the lateral high discharge). This is why the laterals need to be underground.

BUBBLER EMITTERS

The bubblers are small plastic head emitters with a threaded joint. They were originally designed for use on risers above ground for flood irrigation of small ornamental areas. In recent decades they have been used successfully in several countries for the irrigation of fruit trees. They perform well under a wide range of pressures delivering water in the form of a fountain, small stream or tiny umbrella in the vicinity of the emitter. The main performance characteristics are:

- Operating pressure: 1.0–3.0 bars;
- Flow rate (discharge): 100–250 litres/h (adjustable);
- No filtration is required.

There is a wide range of flow rates up to 800 litres/h; this paper presents only low discharge bubblers.

IRRIGATION SCHEDULING

With bubbler irrigation the percentage of the root soil volume wetted is about 80 percent. Thus, there are no restrictions on the way the irrigation programme is prepared. This can be either fixed depletion or fixed interval, taking into consideration the soil water holding capacity, the availability of the irrigation water, the size of flow, etc.

DESIGN CRITERIA AND CONSIDERATIONS

Bubbler irrigation is mainly applied in fruit tree orchards. The most important criteria, apart from the routine design criteria, are the system's special features and characteristics.

Bubbler emitters discharge water on the same spot of ground at high rates. Thus, for a uniform distribution over the basin area, a minimum of land preparation is needed. In sandy soils, the water infiltrates at the point of application and high losses occur due to deep percolation. In fine soils with low infiltration rates, the water ponds and evaporation occurs.

Mature trees always take two bubbler emitters, one on each side, in order to ensure an acceptable uniformity of application. The flow rate per tree is relatively high compared with other micro-irrigation techniques at about 500 litres/h. Thus, the diameter of an 80 m-long lateral for a single row of 13 trees spaced at 6 m intervals should be 50 mm.

The common practice is to have one lateral per two rows of trees with small flexible tubes extended on both sides and connected to the bubblers. In this way, the same size of lateral pipe (50 mm), placed (buried) between two rows can serve 12 trees on each side (24 trees in total) spaced at 6 m intervals with 48 bubbler emitters.

The size of the equipment for the installation should always be able to accommodate the flows required for mature trees.

For longer laterals, pressure compensated bubblers can be used, though this involves higher energy consumption and more expensive higher pressure pipes.

COST

The cost for a complete permanent installation of this system is about US$3 900/ha, of which US$70 are needed for the head control unit, i.e. less than 2 percent. The cost of the pipes (all rigid PVC) is US$1 250 plus US$970 for the trench (excavation and backfilling), i.e. approximately 56 percent of the total cost. The bubblers with the connecting small flexible tubes cost US$980, or 25 percent of the total cost of the system. The cost per tree of the laterals with the bubblers is about US$6.60.

ADVANTAGES

- High irrigation application efficiency, up to 75 percent, resulting in considerable water savings, with absolute control of the irrigation water from the source to the tree basin.
- All the piping network is buried, so there are no field operations problems.
- The technology is simple and no highly sophisticated equipment is used. The system can be operated by unskilled farmers and labourers. No filters or fertilizer injectors are needed.

DISADVANTAGES

- High initial purchase cost.
- Small water flows cannot be used as in other micro-irrigation systems.
- In sandy soils with high infiltration rates, it is difficult to achieve a uniform water distribution over the tree basins.

EXAMPLE DESIGN – Bubbler irrigation with fruit trees

Area and crop

The plot dimensions are 120 *x* 85 m (1.0 ha) with mature guava trees in rows at a spacing of 6 *x* 6 m. There are 20 rows with 14 trees in each row for a total 280 trees. The slope of the plot is 0.5 percent from west to east and from north to south.

Soil, water and climate

Medium texture soil with an infiltration rate of approximately 8 mm/h and a soil available moisture of 150 mm/m depth. The source of water is an existing tube-well with a safe output of 25 m^3/h of suitable quality. The evaporation pan average readings in July are 7 mm/d.

Crop water requirements and irrigation scheduling

The pan reading of 7.0 mm/d multiplied by 0.66 (pan correction factor) gives an ETo of 4.65 mm/d. The crop factor kc is 0.65, thus ETc = 4.65 x 0.65 = 3.0 mm/d. The area shaded by the tree canopy is 70 percent and for calculation purposes it is taken as 82 percent. Therefore, the daily water requirements are: 3.0 x 0.82 = 2.48 mm/d net. With a system application efficiency of 75 percent, the gross daily irrigation requirements are: 2.48 x 100 ÷ 75 = 3.3 mm (33 m^3). If irrigation takes place every ten days, the gross irrigation dosage is: 10 x 33 = 330 m^3.

The maximum permissible irrigation interval in July on a 50 percent moisture depletion for a tree root depth of 0.6 m is: 150 x 0.6 x 0.5 ÷ 3.0 = 15 days. The irrigation frequency depends on many factors and in no case should exceed the maximum permissible irrigation interval.

System layout, performance and hydraulics

The main line also serves as a manifold (feeder) line. The laterals are buried 50 mm rigid PVC pipes laid between every other row of trees with 12 mm flexible tubes rising on both sides and extended to the tree basins with two bubblers in each tree basin (Figures 13.1 and 13.2). The main characteristics of the bubblers are (Tables 13.1 and 13.2):

- flow rate: 225 litres/h at 2.0 bars;
- number of bubblers per lateral: 56 (each lateral irrigates two rows oftrees, i.e. 28 trees);
- lateral discharge: 12 600 litres/h (12.6 m3/h);
- total number of laterals: 10;
- system discharge: 25 m^3/h;
- number of laterals operating simultaneously: 2;
- number of shifts per irrigation: 5;
- operating hours per shift: 2.64 h (2 h 38 min);
- time to complete one irrigation: 13.2 h (13 h 15 min);

TABLE 13.1 - System's operating pressure

	bars
Pressure at the bubbler	2.00
Friction losses in the lateral	0.34
Friction losses in the main line	0.60
Friction losses in the head control unit	0.20
Minor local losses	0.20
Sub-total	**3.34**
Difference in elevation	- 0.10
Total dynamic head	3.24

FIGURE 13.1 - Bubbler in fruit trees. (example design)

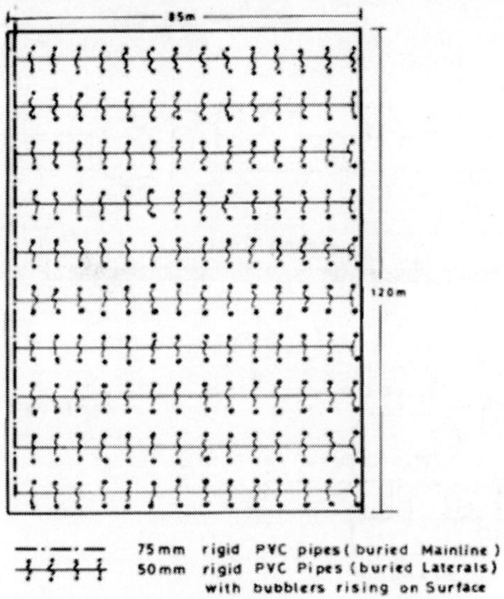

TABLE 13.2 - Equipment required for the system installation

Item	Description	Quantity	Unit price US$	Total price US$
	System distribution network			
1.	75 mm uPVC pipe, 6.0 bars, push-fit joint	120 m	1.90	228.00
2.	50 mm uPVC pipe, 6.0 bars, push-fit joint	850 m	1.20	1020.00
3.	75 mm x 2 in PP clamp saddle	10 pcs	2.25	22.50
4.	50 mm x 1 in PP clamp saddle	140 pcs	1.10	154.00
5.	75 mm x 2 in PP adaptor	1 pcs	9.00	9.00
6.	50 mm x 2 in PP adaptors	10 pcs	4.00	40.00
7.	75 mm PP end plug	1 pc	9.00	9.00
8.	50 mm PP end plug	10 pcs	4.00	40.00
9.	2 in nipple	10 pcs	1.00	10.00
10.	2 in brass shut-off valve	10 pcs	12.00	120.00
11.	1 in PVC tee threaded	140 pcs	1.00	140.00
12.	1 in x ½ in PVC bushing	280 pcs	0.60	168.00
13.	12 mm PP tee barbed	280 pcs	0.30	84.00
14.	12 mm x ½ in PP adaptor barbed	280 pcs	0.25	70.00
15.	12 mm soft flexible PVC tube	1120 m	0.25	280.00
16.	Bubbler set, 225 litres/h at 2.0 bars (adjustable)	560 pcs	0.70	392.00
17.	Valve box, plastic 31 x 50 x 40 cm	10 pcs	20.00	200.00
18.	Trench excavation and backfilling	970 m	1.00	970.00
	Sub-total			**3956.50**
	Head control			
19.	2 ½ in brass check valve	1 pc	15.00	15.00
20.	2 ½ in brass shut-off valve	2 pcs	13.00	26.00
21.	2 ½ in tee (galvanized iron or PVC)	3 pcs	3.50	10.50
22.	2 in nipple	4 pcs	1.00	4.00
23.	1 in single air valve	1 pc	12.00	12.00
	Sub-total			**67.50**
	TOTAL COST			**4024.00**

FIGURE 13.2 - Bubbler irrigation jointing techniques.

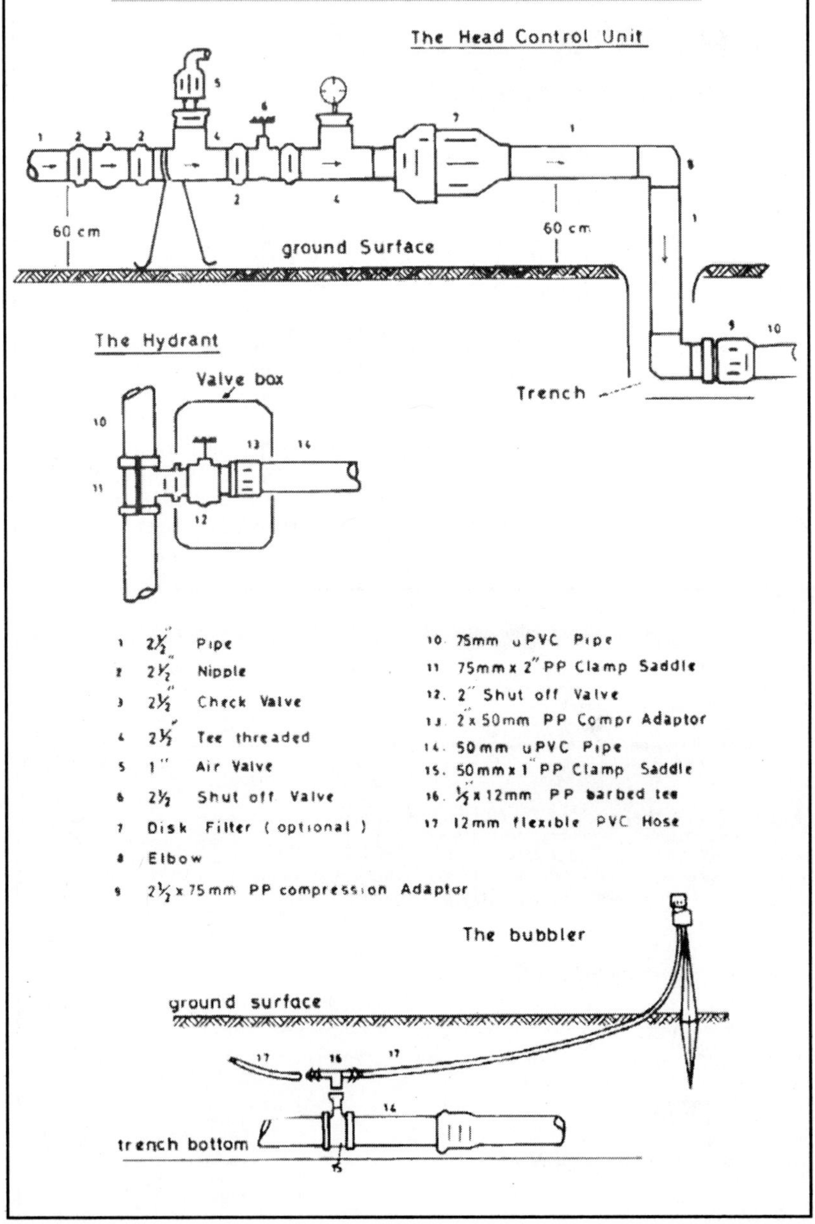

CHAPTER 14: Drip irrigation

INTRODUCTION

In drip irrigation, water is applied to each plant separately in small, frequent, precise quantities through dripper emitters. It is the most advanced irrigation method with the highest application efficiency. The water is delivered continuously in drops at the same point and moves into the soil and wets the root zone vertically by gravity and laterally by capillary action. The planted area is only partially wetted.

In medium-heavy soils of good structure, the lateral movement of the water beneath the surface is greater than in sandy soils (Table 14.1). Moreover, when the discharge rate of the dripper exceeds the soil intake rate and hydraulic conductivity, the water ponds on the surface. This results in the moisture being distributed more laterally rather than vertically. The following water lateral spread values are indicative:

TABLE 14.1 - Type of soil and average radius of water spread laterally with drippers

Type of soil	Average radius of water spread
Light texture	0.30 m
Medium texture	0.65 m
Fine texture	1.20 m

SYSTEM LAYOUT AND COMPONENTS

A complete drip irrigation system consists of a head control unit, main and submain pipelines, hydrants, manifolds and lateral lines with dripper emitters.

Control station (head control unit)

Its features and equipment depend on the system's requirements. Usually, it consists of the shut-off, air and check (non-return) valves, a filtering unit, a fertilizer injector and other smaller accessories.

Main and submain pipelines:

The main and submain pipelines are usually buried, especially when made of rigid PVC.

Hydrants

Fitted on the mains or the submains and equipped with 2–3 inches shut-off valves, they are capable of delivering all or part of the piped water flow to the manifold feeder lines. They are placed in valve boxes for protection.

Manifold (feeder) pipelines

These are usually 50, 63 or 75 mm. Where made of HDPE, they are attached to the hydrants through compression-type, quick release, PP connector fittings and remain on the surface.

Dripper laterals

These are always made of 12–20 mm soft black LDPE, PN 3.0–4.0 bars. They are fitted to the manifolds with small PP connector fittings at fixed positions and laid along the plant rows. They are equipped with closely spaced dripper emitters or emission outlets (Figure 14.1).

FIGURE 14.1 - The mains, manifold and dripper laterals.

In general, the distribution network (mains, submains and manifolds) consists of thermoplastic pipes and fittings (PVC, PE, PP, etc.), PN 6.0 and 10.0 bars. However, for the mains, submains and manifolds, other kind of pipes can also be used, such as quick coupling light steel pipes. In the past, permanently assembled buried rigid PVC pipes were used as mains and submains, with hydrants rising on the surface at desired points. More recently, surface-laid 50–75 mm HDPE pipes, PN 6.0 bars, have been used for the whole distribution network in smallholdings. Larger diameter PE pipes are also available but cost more than rigid PVC pipes of the same size.

The system's pressure ranges from 2.0 to 3.0 bars. Therefore, all drip irrigation systems can be classed as low pressure, localized, solid permanent or seasonal installation systems.

DRIP EMITTERS (DRIPPERS)

The drippers are small-sized emitters made of high quality plastics. They are mounted on small soft PE pipes (hoses) at frequent spaces. Water enters the dripper emitters at approximately 1.0 bar and is delivered at zero pressure in the form of continuous droplets at low rates of 1.0–24 litres/h. Drippers are divided into two main groups according to the way they dissipate energy (pressure):

- orifice type, with tiny flow areas of 0.2–0.35 mm^2;
- long-path type, with relatively larger flow areas of 1–4.5 mm^2.

Both types are manufactured with various mechanisms and principles of operation, such as a vortex diode, a diaphragm or a floating disc for the orifice drippers, and a labyrinthine path, of various shapes, for the long-path ones. All the drippers now available on the market are turbulent flow ones.

Drippers are also characterized by the type of connection to the lateral: on-line, i.e. inserted in the pipe wall by the aid of a punch; or in-line, where the pipe is cut to insert the dripper manually or with a machine.

On-line multi-exit drippers are also available with four to six 'spaghetti' type tube outlets.

Specifications that should be stated by the supplier are:

- dripper discharge (flow rate) at the recommended operating pressure, usually 1.0 bar;
- dripper discharge versus pressure variations and the optimum length of dripper line with different spacing and slopes;
- type of connection;
- filtration requirements;
- coefficient of variation (cv) (the drippers' manufacturing variability).

Dripper emitters which are available as separate items, not built into the pipe, can be referred to as separate source point drippers.

DRIP TAPES

These are thin-walled integral drip lines with emission points spaced 10, 20, 30, 45 cm or any other distance apart, delivering lower quantities of water than the usual drippers at very low pressures, i.e. 0.4–1.0 litres/h at 0.6–1.0 bar. They are integrated drip lines where the drippers are built in the pipe walls at the desired spacing during the manufacturing process. They are ready-made dripper laterals with a very high uniformity of application. Drip tapes are made of LDPE or other soft PE materials in various diameters from 12 to 20 mm and in several wall thicknesses (0.10–1.25 mm). Thanks to a filtration system incorporated inside the tubing, they are less susceptible to mechanical and biological blockages than conventional drippers are.

POROUS-WALL PIPES

These pipes are small-sized (about 16 mm) thin-walled porous flexible hoses made from PE fibres, PVC, ABS or rubber. They permit water and nutrients under low pressure to pass from inside the tube, by transpiration, and irrigate the crops. The porous pipeline discharge is not accurate because the size of the pores varies and is not stable. They are used as lateral drip lines beneath the surface. Their application is limited although they do offer some advantages.

FILTRATION

The filtration of the irrigation water is of major importance for the normal application of this system. The solid content in the water must be removed through effective filtration in order to avoid blockage damage in the drippers. The kind of filtration depends on the impurities contained in the water and the degree of filtration required.

IRRIGATION SCHEDULING

In drip irrigation, the soil volume in the root zone is only partly wetted and the availability of moisture restricted. The soil moisture depletion should not exceed 40 percent of the soil available moisture in the late growing stages of vegetables and fruit trees, and 20–30 percent in the early stages for vegetables. However, in order to obtain higher yields, the common practice is to irrigate every day in the later stages. Proper irrigation scheduling can be arranged by using tensiometers to indicate the soil moisture tension in the root zone. This should range from 10 cbars for light soils to 25 cbars for heavy soils.

DESIGN CRITERIA AND CONSIDERATIONS

Drip irrigation is mainly applied in intensive cultivations planted in rows (vegetables, fruit trees, melons, bananas, papayas, flowers, grapes, etc.). It is not recommended for potatoes, salad leafy vegetables, groundnuts, alfalfa and other dense planted crops, although it can be applied successfully.

The drippers and/or the lateral spacing are directly related to the crop planting spacing. In most vegetable crops, the dripper spacing is identical to the crop planting spacing, i.e. one dripper per plant and one dripper lateral per row of cultivation. With drip tapes there are several emission points per plant in order to ensure a continuous wetted strip along the row. Here, the arrangement is one drip tape per row of crop.

Under drip irrigation, most of the vegetables develop the bulk of their roots in the first 30 cm depth of the soil profile below the emission point. Thus, if both the crop and the emission points along the rows are closely spaced, most of the soil volume can be sufficiently wetted with optimum results.

Where the crop is planted closely in beds, one dripper lateral per two rows might be applied with good results. Other crops planted in double rows (celery, capsicum and hot peppers) are also irrigated by one dripper lateral placed in between the rows.

In widely spaced tree orchards, the dripper spacing differs from that for vegetables. As the soil surface is partially wetted, only a part of the root system is being wetted too. The main consideration is to wet the largest possible volume of soil per tree (root system volume), not less than 35 percent, and at the same time to avoid deep percolation, beyond 50–60 cm, which is the average root depth of fruit trees under drip irrigation. The above percentage corresponds to an area of approximately 10–12 m^2 of soil surface with a tree spacing of 5 *x* 6 m or 6 *x* 6 m. Based on this consideration and the indicative lateral water spread figures, the dripper lateral design arrangements in tree orchards can be as follows:

- Single line per row of trees, with 4–8 drippers at approximately every 0.8–1.2 m along the line;
- Circular layout, or 'loop around the tree'. In this arrangement there is a single line per row and for each tree there is either a smaller extension line with 5–8 drippers around the tree, or a multi-exit dripper with 4–6 small emission tubes extending radially around the tree. The circle diameter can be from 1.2 to 2.2 m. Newly planted trees can have two drippers only on both sides of the trunk, 35–40 cm away from the trunk.
- Double lines per row of plants. This design is applied in banana plantations, with two dripper lines per row, one on each side, set approximately 1.2–1.6 m apart. The drippers along the lines are spaced at 0.7–1.2 m accordingly (Figure 14.2).

FIGURE 14.2 - Double lines with bananas.

COST

The cost for a complete drip irrigation installation is US$4 000–5 000/ha. The cost of the pipes (all tubing, laterals included) is about US$2 000, i.e. 45 percent of the total cost. The head control unit accounts for 30 percent of the total cost.

ADVANTAGES

- Water savings. The planted area is partially wetted with precisely controlled water amounts. Thus, large quantities of irrigation water are saved and the irrigated area can be expanded with the same water supply, resulting in higher income per unit of water.
- Utilization of saline water resources. With drip irrigation, low soil moisture tensions in the root zone can be maintained continuously with frequent applications. The dissolved salts accumulate at the periphery of the wetted soil mass, and the plants can easily obtain the moisture needed. This enables the use of saline water containing more than 3 000 mg/litre TDS, which would be unsuitable for use with other methods.

- Use on marginal fields. Small irregular marginal plots, remote because of land fragmentation with varying topography and shallow soil full of rocks, can be productive under drip irrigation techniques that deliver the required amounts of water and nutrients directly to the plants.
- Low labour operating requirements, reduced cultivation and weed control, and uninterrupted operation are among the other advantages of this irrigation method.

DISADVANTAGES

- High initial purchase cost.
- Good irrigation management is essential for skilled system operation, application of fertigation and maintenance of the head control unit equipment (filters, injectors, etc.).
- Emitter blockages. The first limitation on the successful introduction of drip irrigation techniques in developing countries is mechanical clogging of the emitters because of insufficient filtration of impurities in the irrigation water.

EXAMPLE DESIGN – Drip irrigation in watermelons

Area and crop

The plot dimensions are 120 x 83 m (about 1 ha), planted in the open with watermelons in rows 2.20 m apart and spaced along the rows at 0.5 m. The plot is divided into two parts, each with 54 rows 40.5 m long. There are 81 plants per row. Thus, there are 4 374 plants in each part, i.e. 8 748 plants in the whole plot and 108 plant rows.

Soil, water and climate

Heavy texture soil with low permeability (approximately 6 mm/h) and a high water holding capacity. The source of water is a nearby open water reservoir; it is of good quality but with a high impurity content of organic origin (algae). The crop growing season is from early April to early July; the evaporation pan average maximum readings are 3.3 mm/d in April, 4.64 mm/d in May and 6.13 mm/d in June.

Crop water requirements and irrigation schedule

The maximum irrigation requirements of the watermelons are during the mid-season stage and the yield formation in late May-early June, when the kc value is 1.0. The average reading for the two months is 5.38 mm/d, which multiplied by a correction factor of 0.66 gives an ETo of 3.55 mm/d. As kc = 1.0, ETc = 3.55 mm/d. The system's application efficiency is 90 percent.

Therefore, the daily gross requirements at peak are:

$$3.55 \times 0.90 \div 100 = 3.94 \text{ mm/d}$$
$$3.94 \times 10 \times 1 \text{ ha} = 39.4 \text{ m}^3\text{/d}$$

The irrigation scheduling in late May is not arranged at a fixed depletion of the available soil moisture, but at a fixed interval of one day. Therefore, irrigation takes place every day and the dose is 39.5 m3. At the early stages of the growing season, the irrigation interval ranges from 4 to 2 days.

System layout

The system consists of a head control equipped with a gravel filter and a strainer, a fertilizer injector and a regulating valve. The 63 mm HDPE main line is laid on surface along the middle of the field. On this main line (which also serves as a manifold), there are 54 hydrants 3/4 inch at a spacing of 2.20 m. The laterals, connected to the hydrants, are 16 mm LDPE pipes laid perpendicular to the main line on both sides, one per row of plants. Separate point source drippers are inserted in the laterals at a spacing of 0.5 m, one per plant.

Dripper characteristics:

- on-line: 4 litres/h at 1.0 bar;
- filtration requirements: 160 mesh.

Lateral characteristics:

- pipe: 16 mm LDPE, 4.0 bars PN, length 41 m;
- number of drippers: 81;
- water discharge: 324 litres/h;
- total number of laterals: 108;
- total number of drippers 8 748.

System flow and operation

For the simultaneous operation of all the laterals, the required flow is 35 m³/h. If one irrigation is to be completed in three shifts, the flow of the system is 12 m³/h, a reasonable size of flow for an area of 1 ha. The duration of application per shift at peak demand for an irrigation dosage of 39.5 m³ is 1 h 06 min. The time required to complete one irrigation is 3 h 18 min (Figures 14.3 and 14.4).

Operating pressure

The required pressure for the normal operation of the system is (Tables 14.2 and 14.3):

TABLE 14.2 - System's operating pressure

	bars
Pressure for the drippers	1.00
Friction losses in the dripper lateral	0.10
Friction losses in the main line	0.43
Friction losses in the head control	0.90
Minor local losses	0.22
Total dynamic head	2.65

TABLE 14.3 - Equipment for system installation

Item	Description	Quantity	Unit price US$	Total price US$
	System distribution network			
1.	63 mm HDPE pipe, 6.0 bars	180 m	1.80	324.00
2.	16 mm LDPE pipe, 4.0 bars	4430 m	0.32	2302.60
3.	Drippers, 4 litres/h,1 bar	8750 pcs	0.06	525.00
4.	63 mm x 2 in PP adaptor	1 pc	6.00	6.00
5.	16 mm x ¾ in PP adaptor	108 pcs	0.80	86.40
6.	63 mm PP elbow	1 pc	8.00	8.00
7.	63 mm PP end plug	1 pc	6.00	6.00
8.	63 mm x ¾ in PP clamp saddle	54 pcs	1.30	70.20
9.	¾ in nipple (galvanized iron or PVC)	54 pcs	0.25	13.50
10.	¾ in tee (galvanized iron or PVC)	54 pcs	0.40	21.60
11.	¾ in brass shut-off valves	108 pcs	2.30	248.60
	Sub-total			**3609.70**
	Head control			
12.	2 in brass check valve	1 pc	13.00	13.00
13.	2 in brass shut-off valve	2 pcs	12.00	24.00
14.	¾ in brass shut-off valve	2 pcs	2.30	4.60
15.	2 in tee (galvanized iron or PVC)	3 pcs	2.00	6.00
16.	2 in nipple	4 pcs	1.00	4.00
17.	¾ in nipple	4 pcs	0.25	1.00
18.	1 in air valve	1 pc	12.00	12.00
19.	2 in gravel filter complete	1 pc	600.00	600.00
20.	2 in disk filter, c/with gauges, etc.	1 pc	180.00	180.00
21.	Fertilizer injector complete, up to 150 litres/h	1 pc	500.00	500.00
	Sub-total			**1344.00**
	TOTAL COST			**4953.70**

FIGURE 14.3 - Drip irrigation in watermelons. (Sample design)

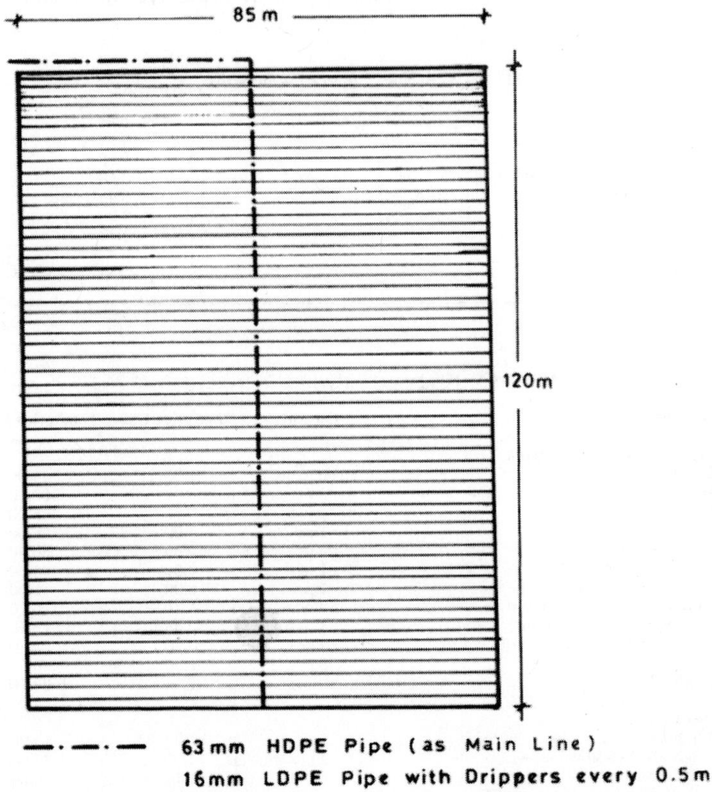

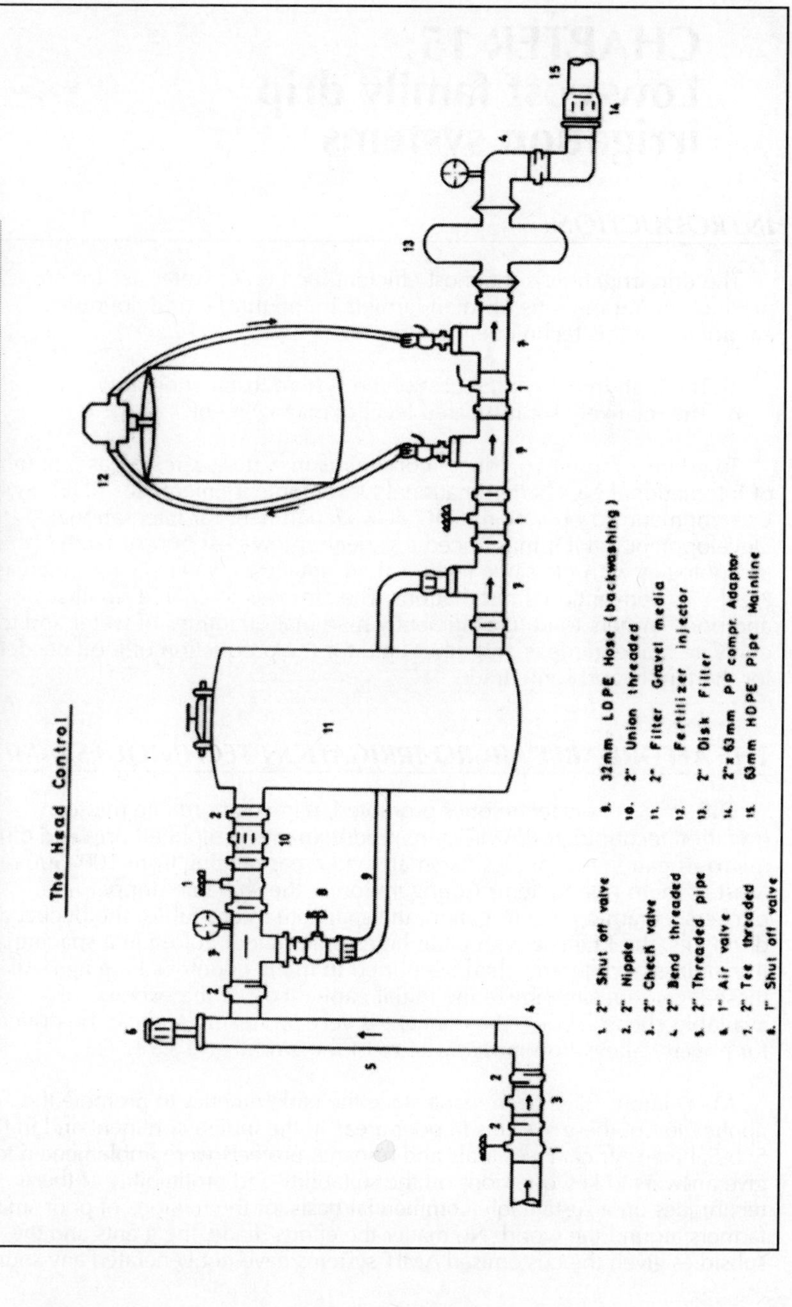

FIGURE 14.4 - Drip irrigation jointing techniques.

CHAPTER 15: Low-cost family drip irrigation systems

INTRODUCTION

The drip irrigation is the most efficient method of water use for crop production. Yet millions of small farmers in the third world countries cannot adopt this technique, due to:

a) The high initial capital cost of the system installation and
b) The relatively sophisticated level of management.

To address these two critical constraints in water scarce areas a number of international NGOs (International Development Enterprises [IDE], Swiss Development Co-operation [SDC], UK Department for International Development [DFID] introduced a variety of low-cost non-surface technologies with the same technical advantages – water savings, increased yield – as conventional installations. The aim was to enable small-scale and poor farmers to utilize efficiently marginal quantities of water and to cultivate home gardens and other land for the production of food needed for the family and some trade.

THE AFFORDABLE MICRO-IRRIGATION TECHNIQUES (AMIT)

The various new techniques promoted, named affordable micro-irrigation technologies (AMIT), are predominantly simplified drip and other micro-irrigation techniques for small land areas ranging from 100–500 m^2. Most of them (see Systems Configuration) – the shiftable dripper line, pitchers (ceramic pots) irrigation, the spaghetti micro-tubes, the bucket and drum kits, the PE hose with plain holes (perforated) at frequent spacing and the multi-exit drippers – had been tried in many countries long ago. All these techniques minimize the initial capital cost at the expense of available cheap labour. The concept is very promising and can be critical for poverty alleviation in many areas of the world.

Many efforts have been made since the early nineties to promote the application of these systems in poor areas in the Indian continent and in the Sub-Saharan Africa. Field trials and research projects were implemented to give answers to key questions on the suitability and profitability of these techniques on a sustainable commercial basis for the millions of poor small farmers around the world. No matter the efforts made, the grants and the subsidies given the customised AMIT systems have not generated any sound interest among the farmers. It must be noted that the Government extension agencies were not in any approach to technology transfer (Figure 15.1).

FIGURE 15.1 - AMIT Configuration of the typical Bucket kit with one 12 mm lateral and spaghetti micro-tubes fitted on either side. Area covered 20 m² (by IDE).

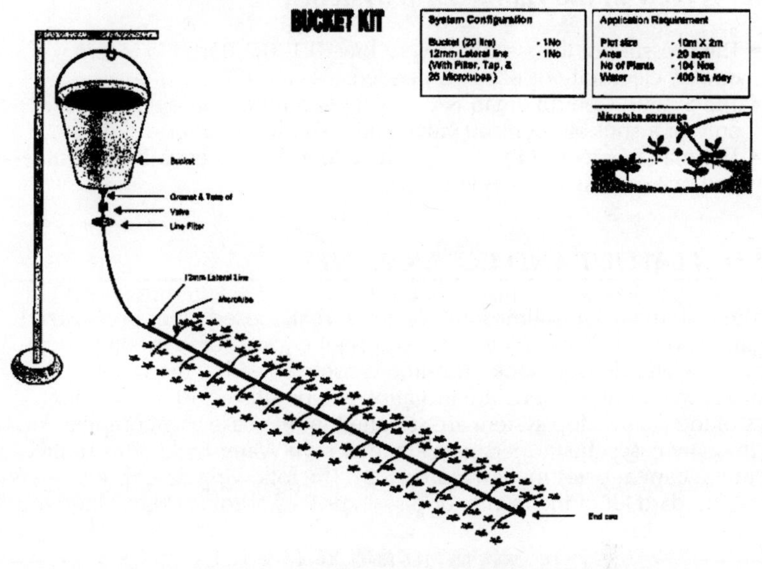

More AMIT configuration are given at the end of the chapter.

Nevertheless, the AMIT systems have drawn the attention of the drip irrigation commercial enterprises. A leading commercial manufacturer in the field of modern irrigation developed the so-called family drip system (FDS). It is an improvement of the Drum kit. Other big companies have also included this type of system in their range of products under various names, e.g. "Easy Drip" etc.

THE FAMILY DRIP SYSTEM

This system is a complete drip irrigation unit; it operates by water gravity from a tank placed at 1–1.5 meters high. It is a closed piped gravity system, localized method, and solid seasonal installation, for growing vegetables, flowers and other horticultural crops on flat or minor slope land. It does not necessarily need any external power for normal operation. The average system flow is around 1 m^3/h. Developed especially for family farming as a single production unit, the typical system sizes range from 500–1000 m^2. It is ideal for rural conditions and small-scale agricultural production in rural areas with water shortage and limited supply. It can be installed in greenhouses, low-tunnels and in the open and in the back yard of village houses.

What is New in the Family Drip Systems?

- The pressure of the system is very low (0.1–0.2 Bar). No external power, electricity or other, is needed.
- The System Control Head is very simple and inexpensive. It consists only of a shut off (control) valve and a small screen or disk filter.
- The management of the Family kit systems is very easy. No skilfulness and/or expertise are needed.

SYSTEM LAYOUT AND COMPONENTS

The system layout is almost the same as in all closed pipes pressurized irrigation systems. A complete Family drip kit consists of four component parts - a "water delivery tank", "a simple control head", "the water distribution pipelines" and "the irrigation dripper lines". All component parts of the Family drip system are supplied in one case as a complete kit. Yet in many cases the farmers provide their own Water Tanks. This reduces the initial capital investment (Figure 15.2). The following description refers to the standard kit of the Family drip systems for approx. 500 m^2 land area.

FIGURE 15.2 - The Family Drip System.

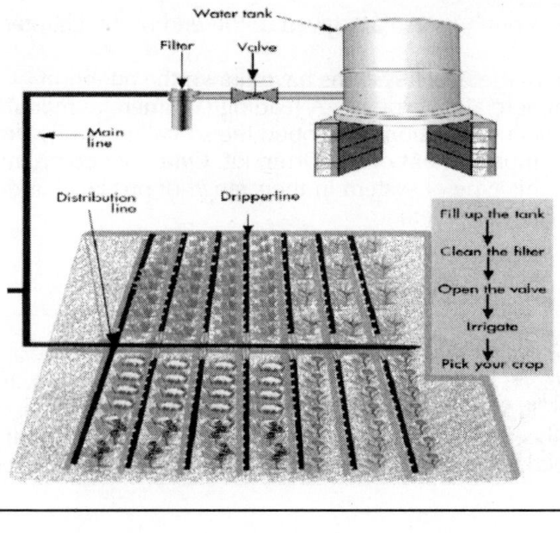

The water tank

The water tank is of 300–3000 liters capacity. It can be a plastic barrel, one, two r more, an iron drum, a self-made with concrete or any other suitable tank. It is always placed 1–1.5 m above ground level, so that the system will have enough gravity for water pressure. Filling the tank method depends on the particular situation, by buckets, by a hand pump or a treadle pump or by a small diesel pump. The source of the water can be canals, or ditches, shallow wells, rivers and boreholes. The tank is always covered, to protect the water from the various impurities and the sunlight and prevent algae development. It has a drain tap at the bottom for frequent flush out and cleaning from suspended solid particles. A plastic tank outlet 1 inch threaded male is fixed on the tank at least 5 cm above the bottom. This fitting is necessary to be included in the system's parts, with the tank or separately (Figure 15.3).

FIGURE 15.3 - Locally made Tank for FDS.

The control head

The control head consists of a "control" plastic shut off valve 1 inch (gate or ball valve), and a 1 inch plastic (PP) line filter, screen or disk type, 120 mesh. The control valve is connected directly to the tank outlet and then the filter is installed.

The water pipelines

The water pipelines 20 mm, 25 mm, or 32 mm size black PE pipes PN 2.5 Bars, 50 m to 75 m long according to the size and/or the shape of the land. They serve as main and as laterals laid along the borderline of the field. The very few pipe connector fittings needed are polypropylene quick release compression type. Small size plastic start connectors (adaptors), barbed type, are inserted to the lateral pipelines at spacing according to the rows of the crop for feeding the dripper lines.

The dripper lines

The dripper lines are integrated drip-lines 8 mm or 12 mm size, made of a mixture of Low and Linear density PE to ISO 9260/61, with wall thickness 0.6 mm to 0.9 mm (24 mil–35 mil). The built-in point source dripper emitters are labyrinth type with wide water passages and protection filter to avoid clogging. They are normally spaced at 30 cm frequent intervals. The dripper discharge ranges from 0.5–0.7 liters/h at 1meter pressure head and from 0.65–0.95 liters/h at 2 m pressure head, depending on the dripper. The average recommended length of the dripper line at 1 meter pressure head, on relatively level ground, is 12–15 m for the 8 mm size and 25–30 m for the 12 mm sizes. They are laid along the rows and push-fit connected to the water pipelines. The total length of the dripper lines for approximately 500 m^2 lands is around 500 m.

Very often the use of a small pump is essential when there is a number of neighboring farmers with water rights from the same well. In such cases larger tanks for each individual farmer are recommended.

IRRIGATION SCHEDULING

Irrigation requirements

One of the practical difficulties in the concept of the AMT is how much water and when to apply it. The total irrigation water requirements of a certain crop are always the same and depend on the climatic conditions, the growing season and the method of irrigation application. They can be calculated following FAO methodology. The considerable savings in water with drip and other micro-irrigation are mainly due to the high application efficiencies of the techniques. The total irrigation water needs of most of the annual field crops (vegetables and melons) vary from 300 mm to 650 mm, depending mainly on the length and the time of the growing period.

Number and frequency of irrigations

In the calculation of the water needs the area factor is replaced with the number of plants. Area covered is the number of plants multiplied by the planting spacing. As in the conventional drip irrigation in the Family drip systems the irrigated area and volume of soil is partly wetted and the availability of moisture in the soil is restricted. Then small frequent applications are needed in all seasonal cultivations as well as in perennials. This must always be explained to the farmers who are using drip and other micro-irrigation techniques.

The minimum number of irrigations per season in vegetables can be from 40 (water melons) to 75 (tomatoes) according to the crop and the length of growing period. This varies from 75 days (fresh beans, green onions, squashes) to 170 days (tomatoes). Pulse irrigation may raise the number of irrigations up to 120. The frequency of irrigation normally may be from 1–3 days with an average water dosage of 0.5–1 litre/day/plant at the first stage of growth. During the yield formation the dosage increases to meet the requirements of 1–2 litres/day/plant. Later the daily needs increase up to 3–6 litres/day/plant and at late season, sometimes, 5–7 litres/day/plant. Daily applications are needed. During the stage of harvest the requirements decrease by 10–20 percent. (The higher needs mentioned are for watermelons). **Irrigation Schedules for all crops will be prepared on the spot.**

DESIGN CRITERIA AND CONSIDERATIONS

Area, size and shape

The area can be any agricultural land, planted with field crops in rows of short lengths from 12 to 24 meters located in the rural areas on the mountains or in the plains. The size can be from 250 m^2 to 1000 m^2. The standard Family drip systems are designed for 500 m^2, however smaller or larger fields can be irrigated depending on the crop planting spacing. **The shape should be of normal rectangular or square shape.**

Topography and type of soil

These irrigation systems operate at very low pressures, so flat level lands are recommended and/or uniform sloping fields with slopes < 0.5 percent. The soil can be of any texture, preferably medium and/or fine and with an infiltration rate < 20 mm/hour. Very light sandy soils with high permeability are not recommended.

Water availability

The source of water can be a small hand-dug well, a tub, small seasonal rivers, ponds and ditches or any other which can fill the system's water tank regularly. Hand-pumps, treadle pumps and small diesel pumps are often used to fill the tank regularly. The designed flow of the Family drip irrigation systems offered is around 1.1 m^3/h at 1.5 m head.

Water quality

The water should be, as clean as possible although there is a complete filtration system. Chemically it must be of normal pH 6.5 8.4, with low to medium salinity, low sodium hazard and toxicity problems caused by bicarbonates, nitrates or boron. TDS can be from 500 up to 2 000 mg/l (ppm), SAR < 12, RSC < 1.25 meq/l. Most of the seasonal vegetables and the melons are relatively tolerant to boron content from 1.5 2.0 mg/l. But strawberries and beans are sensitive, so the boron content should not exceed 0.7 mg/l and the total salinity not to exceed the 1.5 dS/m (1 000mg/l).

Kind of crops

All data on the field crops, which can be grown under drip irrigation techniques in an area, the growing season in winter and summer time, planting spacing, etc. are known to the local agronomists and the farmers. The table "Useful data on common seasonal crops" is very important for the design of the FDS and any other AMIT. The local agronomists and farmers can give data for any field crop (Figure 15.4).

Special consideration

Duration of irrigation application

In all micro-irrigation systems the irrigation dosage is applied simultaneously to a larger area, but at lower rates per unit of area (plant) in comparison with the other techniques and methods of irrigation. In the

FIGURE 15.4 - FDS dripper lines in China.

FDS the discharge from the dripper emitters is very low, around 0.65 l/h. At the last stage of crop growth many plants need up to 7–8 litres per day, so the duration of irrigation should be almost 5 hours for 1 000 m² land. The duration of irrigation application is directly related with the system flow, the size of the tank, and the number of fillings and the availability of water. E.g. for a drum of 200 litres capacity with a system flow of 1.1 m³/h for a daily application of 2 750 litres of water on 500 m² cultivated land, the duration of application would be 2.5 hours and the number of fillings around 14. This factor should be considered seriously in the planning and designing. Partial water application (pulse irrigation) is recommended during high water demands.

The Installation of the FDS

During installation the connection between the various components and fittings should be made properly, so that water will not leak. Leaking may affect seriously the performance of the system. Teflon plastic tape is used in all threaded connections. In FDS, like in all pressurized systems, when the installation is completed it is necessary to flash the system. First the water pipes are flashed, then the dripper lines are attached to the start connectors and the valve is open again. When the dripper lines are flashed too then the system is ready for operation.

COST

The initial capital cost for a complete Family drip piping system unit without the Tank is US$100–120 for a land area of around 500 m². The price of a plastic tank of 300–500 litres capacity is almost the same with the piping system. In the many cases where the farmers use their own self-made tanks the cost is less. Two system units at a cost of US$200–240 may cover a land area of 1 000 m². In case of the need of a water supply small pump for a group of farmers the extra cost is divided by the number of users (Table 15.1).

TABLE 15.1 - Useful data on common seasonal crops in the open (all values approximate)

Crop	Planting (transplant.)	Growing Period (days)	Harvest	Plant. Spacing (cm) Along rows	Plant. Spacing (cm) Between rows	No. of plants for 100 m²	Minimum No. of irrigations
Beans (trell., non-trell.)	Feb.–Mar. &Aug.	75–90	May–Nov. average	15–22 (5–10)	90 (45–60)	500–700 2 800	60
Cabbage	Aug.–Apr.	120–140	Year around	30–45	45–75	300–700	20
Celery	May–Oct.		Aug.–May	25–30	30–45		25
Cucumbers	Mar.–May Aug.–Sep.	105–130	May–Aug. Oct.–Nov.	30–50	130–160	120–240	55
Eggplants	Apr.–May	125–140	Jun.–Dec.	60–75	90	150	65
Lettuce	Sep.–May	70–90	Year around	15–30	30–45	750–2 200	20
Onions green	Sep.–Feb.	70	Nov.–May	5	15		15
Peppers	Apr.–May	120–180	Jul.–Nov.	45–60	75	200–300	60
Squashes	Mar.–Sep.	90–120	May–Nov.	60	120–140	100–140	50
Tomatoes (trellised)	Mar.–Sep.	135–190	Jun.–Dec.	45–60	90–150	180–250	65
Tomatoes (non-trell.)	Mar.–Sep.	135–180	Jun.–Dec.	60–75	150	100	65
Watermelons	Mar.–May	90–120	Jun.–Sep.	50–75	180–240	60–90	50

Note:

The growing seasons and periods of vegetables and annual field crops vary according to the local climatic conditions. In general the growing period of most of the seasonal crops is from 90 days to 170 days. The above indicative figures cover most of the arid and semi-arid regions of the N. Hemisphere.

This cost is significantly lower than the cost for a conventional high-level management drip irrigation system with the same coverage. In conventional sophisticate systems the Control Head accounts from 30 percent to 45 percent the total initial capital cost, whilst in Family drip systems there is no need for such investment and the Control Head comprises only of a valve 1 inch and a line filter of same size. At this total cost the Family Drip Systems can be classified as Low-cost irrigation systems.

ADVANTAGES

The price is very low, it is easily installed and operated and has all the advantages of drip irrigation, when correctly use and maintained. Fertigation can be applied through the system by diluting the soluble fertilizers in the tank of the system. The expenses of maintenance and the total cost of irrigation are reduced significantly.

There is no energy consumption. The pressure required for the system operation is very low (1–2 m head of water) as in gravity systems. In conventional drip installations the initial pressure at the Control Head is 3.0–3.5 Bars.

DISADVANTAGES

The lack of training and knowledge among the farmers on the agronomic aspects create negative results. The simplicity of the system in many cases turns into a negative factor as illiterate farmers in many countries do not maintain the dripper lines and the system's life is shortened to one or two seasons only.

EXAMPLE DESIGNS – Family drip systems in tomatoes (trellised)

Area and crop

The plot dimensions are 20 x 25 m (500 m^2) planted in the open with tomatoes in rows 1.50 m apart and spaced along the rows 0.60 m. The plot is divided along into two parts, each one having 13 rows 12.5 m long. There are 21 plants per row. So there are 273 plants in every part, i.e. 546 plants in the whole plot and 26 plant rows. Planting is planned to be done in late July/early August and the growing period to be extended up to early next year 140 days approx.). Irrigation stops in late November, but harvesting continues until February.

Soil and water

Medium texture soil with permeability around 12 mm/h and relatively good water holding capacity. The source of water is a nearby shallow hand-dug well equipped with a small pumping unit; it is of good quality but with low impurities content. Filling of the system's tank is done directly from the well with the use of a PE hose.

Crop water requirements and irrigation schedule

The total irrigation requirements of the tomatoes are around 650 mm. The calculation is made following the FAO methodologies. The irrigation scheduling in this system is not arranged at a fixed moisture depletion of the available soil moisture, but at fixed interval of one day, two and three days. So irrigation takes place frequently and the dose varies according to the stage of growth of the crop. At the last, the harvest stage, the irrigation depends on the effective rainfall and the price of the tomatoes.

In this example design the crop is planted in mid summer at the highest Eto values. It is an extreme case as example, however very common in practice. As already mentioned above, *"The frequency of irrigation normally may be from 2–5 days with an average water dosage of 0.5–1 litre/day/plant at the first stage of growth. During the yield formation the dosage increases to meet the requirements of 1–2 litres/day/plant. Later the daily needs increase up to 3–6 litres/day/plant and at late season, sometimes, 5–7 litres/day/plant. Daily applications are needed."*

In the example case the water demand is very high in the early stages of the crop, although the crop ground cover is limited and increases as the crop is entering the mid and late season stages. (During the crop development and the mid-season stages and the kc value is 1.0, whilst in the early stages is 0.45–0.75. The systems application efficiency is 90 percent. In this example the following irrigation program can be used as a guideline (Table 15.2 and Figure 15.5):

TABLE 15.2 - The irrigation program of the case example

Growing period (stages)	Irrigation demand (mm)	Interval (days)	Dose litres/ plant	Irrigation dose (m^3)	Number of irrigations	Total water applied (m^3) per 500 m^2
30 Jul.–15 Aug.	90	1	5.5	3.0	15	45
16 Aug.–10 Sep.	155	1	5.5	3.0	25	75
11 Sep.–10 Oct.	155	2	9	5.170	15	76
11 Oct.–15 Dec.	210	3	8	4.722	22	104
16 Dec.–15 Jan.	50	Effective rainfall equivalent to ETc				25
TOTAL	**660**	–	–	–	**77**	**326**

FIGURE 15.5 - The FDS system layout.

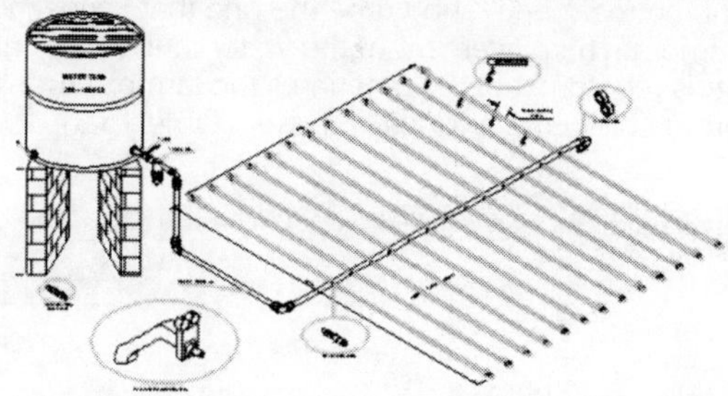

The layout of the system (description and characteristics), performance and hydraulics

- Area covered: 0.5 ha (average) (Plot dimensions 25m x 20 m).
- Type of Systems: Low-pressure, drip irrigation, solid systems installed at fixed seasonal positions.
- System layout: The 25 mm LDPE water pipeline crosses the area along the 20 m dimension with dripper lines one per row of plants fitted perpendicular on both sides.
- System's Component parts: a) Water tank (200 l), b) 1 inch control valve, c) 1 inch Filter, d) 25 mm LDPE main Pipelines, e) 8 mm Drip lines.
- System flow and pressure: 1 100 l/h at 1.5 m head approximately.
- Number of drip lines: 40
- Length of Dripper lines: 12 m
- Total length of dripper lines: 480 m
- Dripper discharge: 0.5–0.65 l/h at 1 m head and 0.65–0.80 l/h at 2 m head.
- Dripper spacing along the line: 30 cm
- Spacing between the lines: 1.5 m
- Total number of drippers: 1 600
- Entire area irrigated simultaneously.
- Number of shifts to fulfil one irrigation: 1 shift

Area covered (irrigated) depends on the kind of crop and especially the crop planting spacing. There are various layouts of the system depending on the size and the shape of the field. The drippers' frequent short spacing along the line ensures a continuous wetted strip along the plants rows, as in furrows.

Hydraulics of the system

The pressure required for the normal operation of the dripper lines is 1–1.5 m head of water. The loss of head due to friction in the 25 mm LDPE main pipelines with 1.1 m/h flow is around 6 percent for pipes without outlets and 2.25 percent when water is distributed en route. So, the layout of the system and especially of the main lines should be the one that keeps these losses of head to the minimum. The placement of the water tank in the middle of the plot or next to it, is recommended. Sometimes the farmers arrange the tank at a higher position to compensate for these losses (Table 15.3).

TABLE 15.3 - List of equipment for system installation (Low-cost family drip irrigation system)

Item	Description	Quantity	Cost US$
1.	LDPE pipe 25 mm	45 m	
2.	PP QR elbow 25 mm x 1 in (F)	1 pc	
3.	PP QR elbow 25 mm	1 pc	
4.	PP barbed start connector 8 mm	45 pcs	
5.	PP barbed start connector 8 mm blind	10 pcs	
6.	Plastic shut off valve 1 in (F)	1 pc	
7.	Plastic screen filter 1 in (M) 120 mesh	1 pc	
8.	Plastic tank outlet 1 in (M)	1 pc	
9.	LPDE drip line 8 mm with built-in point source drippers at 30 cm frequent intervals	500 m	
	TOTAL COST		**100**

Note:
The cost for the tank is not included.

VARIOUS AMIT SYSTEMS CONFIGURATION (AFTER IDE)

FIGURE 15.6 - Drum kit using a 16 mm sub-main five 12 mm laterals with spaghetti micro-tubes in 120 m² area.

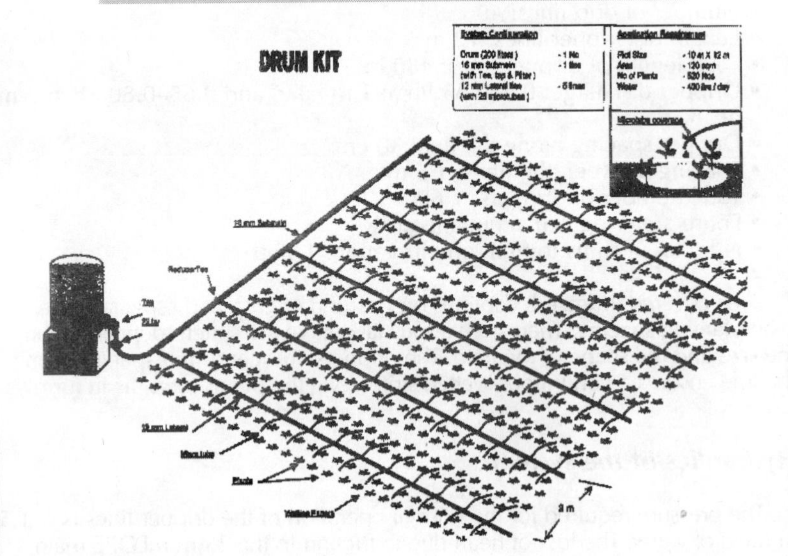

FIGURE 15.7 - System with dripper lines using standard sized holes as dripper emitters covered with plastic sleeves to control water discharge.

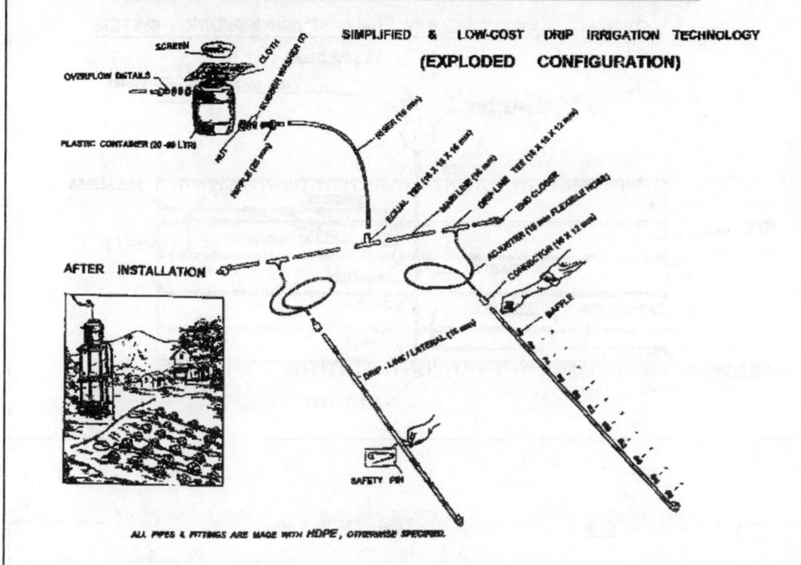

FIGURE 15.8 - Microsprinkler kit in 250 m.

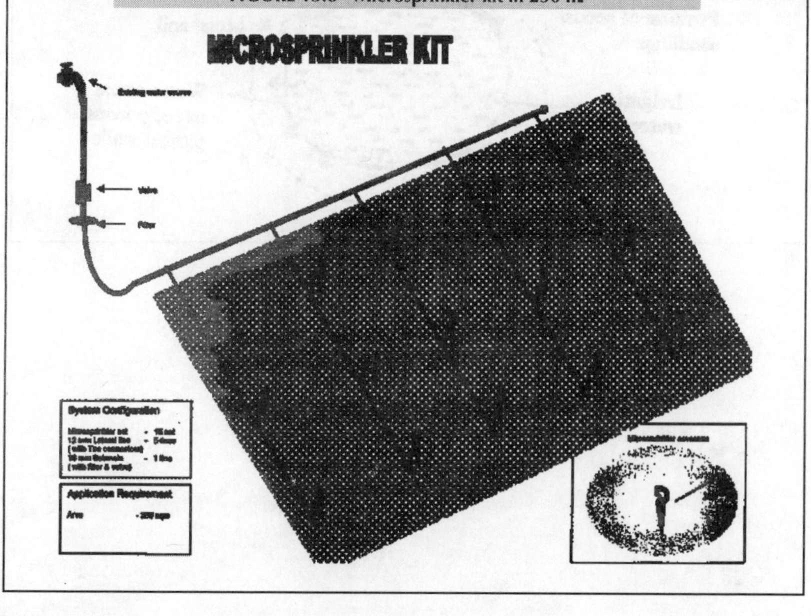

FIGURE 15.9 - Shiftable dripper lines system.

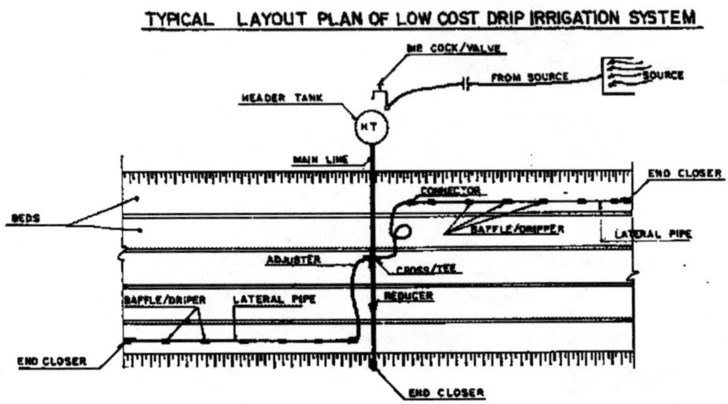

FIGURE 15.10 - Pitcher irrigation.

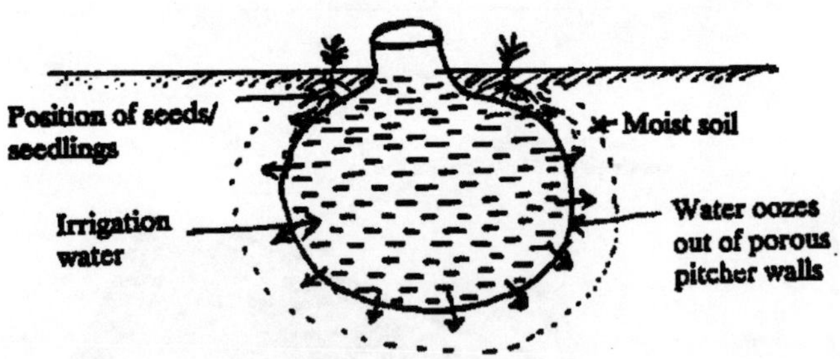

CHAPTER 16: Fertigation

INTRODUCTION

In micro-irrigation, fertilizers can be applied through the system with the irrigation water directly to the region where most of the plants roots develop. This process is called fertigation and it is done with the aid of special fertilizer apparatus (injectors) installed at the head control unit of the system, before the filter. The element most commonly applied is nitrogen. However, applications of phosphorous and potassium are common for vegetables. Fertigation is a necessity in drip irrigation, though not in the other micro-irrigation installations, although it is highly recommended and easily performed.

FERTILIZER INJECTORS

Several techniques have been developed for applying fertilizers through the irrigation systems and many types of injectors are available on the market. There are two main techniques: the ordinary closed tank; and the injector pump. Both systems are operated by the system's water pressure. The injector pumps are mainly either Venturi type or piston pumps. The closed tanks are always installed on a bypass line, while the piston pumps can be installed either in-line or on a bypass line.

❶ **Fertilizer (closed) tank.** This is a cylindrical, epoxy coated, pressurized metal tank, resistant to the system's pressure, and connected as a bypass to the supply pipe of the head control. It operates by differential pressure created by a partially closed valve, placed on the pipeline between the inlet and the outlet of the tank. Part of the flow is diverted to the tank entering at the bottom. It mixes with the fertilizer solution and the dilution is ejected into the system. The dilution ratio and the rate of injection are not constant. The concentration of fertilizer is high at the beginning and very low at the end of the operation. However, this apparatus is still in service on a very small scale in some countries because of its low cost and easy manufacture (Figure 16.1).

❷ **Venturi type.** This is based on the principle of the Venturi tube. A pressure difference is needed between the inlet and the outlet of the injector. Therefore, it is installed on a bypass arrangement placed on an open container with the fertilizer solution. The rate of injection is very sensitive to pressure variations, and small pressure regulators are sometimes needed for a constant ejection. Friction losses are approximately 1.0 bar. The injectors are made of plastic in sizes from ¾ to 2 inches and with injection rates of 40–2 000 litres/h. They are relatively cheap compared to other injectors.

3. **Piston pump.** This type of injector is powered by the water pressure of the system and can be installed directly on the supply line and not on a bypass line. The system's flow activates the pistons and the injector is operated, ejecting the fertilizer solution from a container, while maintaining a constant rate of injection. The rate varies from 9 to 2 500 litres/h depending on the pressure of the system and it can be adjusted by small regulators. Made of durable plastic material, these injectors are available in various models and sizes. They are more expensive than the Venturi-type injectors.

FIGURE 16.1 - The Fertiliser Injectors.

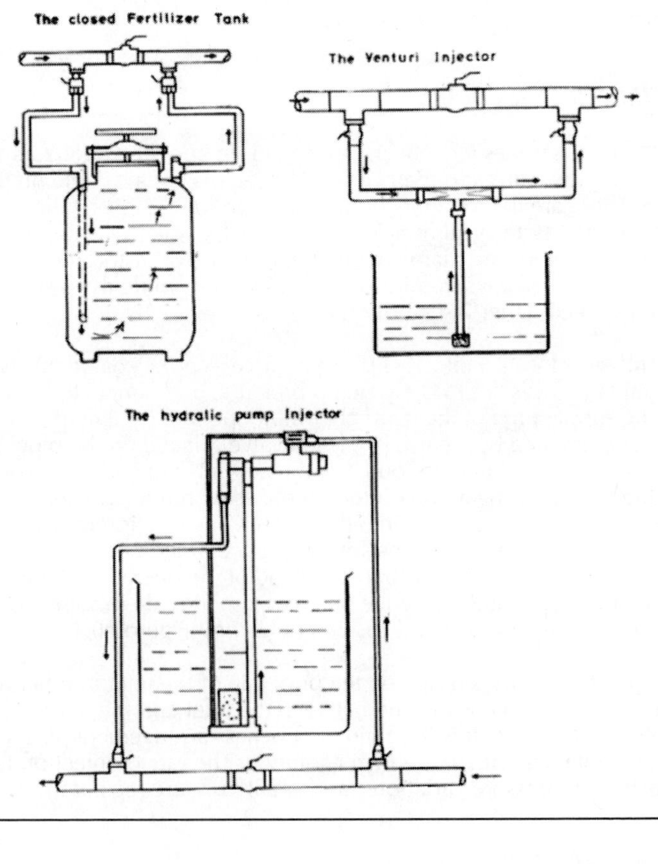

FERTILIZER APPLICATION

The fertilizer solution in liquid form is fed into the system at low rates repeatedly, on a continuous basis, during irrigation. The flow rate of the injector should be such that the calculated amount of solution is supplied at a constant rate during the irrigation cycle, i.e. starting fertigation right after the system starts operation and finishing a few minutes before the operation ends. Regarding the choice of the fertilizers, apart from the amount and the kind, other parameters need to be considered, such as solubility, acidity, compatibility and cost.

Solubility

The fertilizer stock solution should always be dissolved in a separate container and then poured into the suction tank (Figure 16.2). The types of fertilizer should be highly soluble and when dissolved in water must not form scums or sediments which might cause emitter clogging problems. The solution should always be agitated, well stirred and any sludge deposited in the bottom of the tank should be periodically removed. The injector suction pipe should not rest on the bottom of the tank. Hot water helps dissolve dry fertilizers. Their degree of solubility varies according to the type and the country of origin. Potassium nitrate (13-0-46) seems to have a low solubility of approximately 1:8, i.e. 1 kg of dry fertilizer in 8 litres of water. The solubility of potassium chloride (0-0-62) is 1:3, while ammonium nitrate (34-0-0) and calcium nitrate (15.5-0-0) have a high solubility of approximately 1:1. Dry phosphorous fertilizers have a lower solubility than nitrates at about 1:2.5.

Acidity

The acidity produced by the several forms of nitrogen varies from type to type and is greatly affected by the kind of irrigation water and the type of soil. At least one check on the soil pH should be carried out at the beginning of the season and one at the end. Furthermore, a complete ionic analysis of the water is necessary.

Quantity

A simple method for calculating the amount of fertilizer required for fertigation is to divide the annual application by the number of irrigations. Various recipes have been developed in different countries based on the conventional nutrition dosages. The total quantity of fertilizers applied is also related to the length of the growing season and the irrigation requirements.

Table 16.1 presents some of the recipes applied in Cyprus for fertigation on a continuous basis, at a constant rate and feeding, during irrigation.

TABLE 16.1 - Net concentration of fertilizers in ppm (mg/litre, or net fertilizer g/m³ irrigation water)

Crop	N	P	K
Citrus	50	12	15
Bananas	50	15	40
Tomatoes	180	50	250
Cucumbers	200	50	200
Bell peppers	170	60	200
Cabbage	100	60	200
Onions	100	50	150
Squashes	200	50	200
Potatoes	150	50	180
Groundnuts	120	50	200
Watermelons	150	50	150

Note:
The above recipes vary in accordance with the fertilizer reserves in the soil.

FIGURE 16.2 - Preparing the fertilizer solution.

The above recipes are recommended for irrigation water with very low salinity. As a rule of thumb for average quality water, the maximum fertilizer concentration, which is added to the irrigation total salinity, should have an EC of about 0.5 dS/m. For higher concentrations, the salinity level in the soil root zone must be checked frequently and the application adjusted according to the soil test results.

EXAMPLE – Fertigation with vegetables

- Crop: Tomatoes;
- Concentration of NPK fertilizers: 180-50-250;
- Type of fertilizers available: Ammonium nitrate (33.5-0-0) NH_4NO_3; Diammonium phosphate DAP (16-48- 0); $(NH_4)2HPO_4$; Potassium chloride (0-0-60) K_2O;
- System flow: 23 m^3/h;
- Irrigation dosage: 18 m^3;
- Duration of application: 1.5 hours.

Phosphate and potassium are given in oxides, therefore they are converted into P and K elements by multiplying by 0.4364 and 0.8302 respectively.

Calculation of the amounts of fertilizers needed in grams per cubic metre of water:

$$K = 250 \times 100 \div (60 \times 0.8302) = 0.502 \text{ kg } K_2O$$

$$P = 50 \times 100 \div (48 \times 0.4364) = 0.239 \text{ kg } (NH_4)2HPO_4$$

This amount also provides 38 g of N.

$$N = (180-38) \times 100 \div 33.5 = 0.424 \text{ kg } NH_4NO_3$$

Thus, for 18 m^3 of water, which is the irrigation dosage, the exact quantities are:

$$0.502 \text{ kg} \times 18 = 9.036 \text{ kg } K_2O$$

$$0.239 \text{ kg} \times 18 = 4.30 \text{ kg } (NH_4)2HPO_4$$

$$0.424 \text{ kg} \times 18 = 7.63 \text{ kg } NH_4NO_3$$

The amount of water needed for the dilution of the above quantity of fertilizers is estimated by taking into account the solubility of the fertilizers:

9.036 kg K_2O*x* 3 litres	27.00 litres
4.30 kg Ca (H_2PO_4)*x* 2.5 litres	10.75 litres
7.63 kg NH_4NO_3 *x* 1 litre	7.63 litres
Minimum amount of water needed	45.00 litres

If the fertilizers are diluted in 60 litres of water and the duration of the irrigation is 1.5 h (1 h 30 min), then the injection rate should be about 40–45 litres/h in order to complete the fertigation in approximately 1 h 25 min.

CHAPTER 17: Low-cost pipes distribution system

INTRODUCTION

Pump irrigation covers the majority of the irrigated lands in the developing countries of the arid and semi-arid regions. The irrigation water taken from the aquifer or the surface run off ponds, lakes, rivers and dams, is pumped to the fields through conventional (earth) ditches or lined canals, resulting into tremendous losses from seepage and evaporation, deep percolation and canal leakage.

Studies from many countries show an average of 33 percent water losses during conveyance through a 100 m conventional channel. The field irrigation methods are the traditional surface gravity - furrow, basin, border etc., with field application efficiencies of 60–70 percent, i.e. additional water losses of about 20–27 percent of the total. Then the overall irrigation efficiency ranges from 40 pecent to 47 percent approximately.

The solution to the problem is the closed pipes improved irrigation techniques. The huge gap between the water wasteful open surface irrigation practices and the highly efficient improved irrigation techniques

FIGURE 17.1 - Irrigating young trees with the PDS.

can be eliminated with the implementation of the Pipe Distribution Irrigation System (PDS). This irrigation technique had been extensively applied in Cyprus in the early sixties and in Yemen (Tihama) in early eighties with big success. It is actually the replacement of the open channels with a properly designed closed piping network to convey and distribute the irrigation water up to the field plots without any losses. It is a simple technology with the minimum cost, which in one night may raise the overall irrigation efficiency from 40 percent up to 77 percent. The PDS is a closed pipe surface irrigation technique and it is classified as a low-medium pressure system with solid permanent installation (Figure 17.1).

SYSTEM LAYOUT AND COMPONENTS

The basic layout of the PDS consists of a simple control head, a pipe distribution network and the hydrants.

The ***control head***includes the necessary regulating valves (shut off, check valve, air valve) placed on a piece of galvanized steel threaded pipe, 60 cm above ground surface, with tee outlet for taps and pressure gauge. This arrangement, at a later stage, can easily be converted to a more sophisticated control unit, suitable for micro-irrigation systems.

The ***main and sub-main pipelines***(distribution network) can be of rigid PVC, 90–160 mm DN, 4–6 Bars PN, buried underground. On hilly areas other kind of pipes are used on surface ground, such as the flexible black Polyethylene (HDPE), the Quick Coupler light steel or the galvanized steel threaded pipes. The latter is used only up to the 3 in size, because of the high cost.

The ***hydrants*** are rising on surface equipped wit a shut off valve (gate valve) capable to deliver part of the systems flow or the whole of it to the manifold open ditches. At a later stage portable lightweight pipes (quick couplers aluminum, light steel, lay-flat hose, black polyethylene, etc.) can be attached to the hydrants, replacing the manifold open ditches, for the final delivery. From the hydrants the irrigation water is discharged directly to the manifold open earth channels for diversion to the furrows, the basins or the strip borders.

DESIGN CRITERIA AND CONSIDERATIONS

The pipes distribution system combines both the features of the open surface methods and the pressurized closed pipe techniques. The design criteria and the parameters are too many as compared to the simplicity of the installation (Figure 17.2). The topography of the area (shape, slope, etc.), the type of soil, the size of flow and the method of water delivery to the crop (furrow, basin, border or other) should be carefully examined. The take-off hydrants must be placed at the highest points of the field plots and at the right distances to enable efficient practices of the gravity irrigation techniques through the manifold ditches.

"The most important criterion to be considered during the design is the possibility for future extension of the network for the adoption of any other low-medium improved irrigation system, such as sprinkling, drip, spitters, etc., with the minimum expenses. Then the careful design of a flexible skeleton-piping network, suitable to serve all methods of irrigation and water delivery techniques is of major importance".

FIGURE 17.2 - Installation of the pipes.

In this kind of installations the size of the pipes is not reduced at the secondary and tertiary branches (sub-mains, etc.), but remains the same all along the network (Figure 17.3). Thus the system is capable to deliver its total flow at any point of the farm, through each separate hydrant. This results into some extra cost for the pipes. There are no limitations on the kind of the pipes and fittings to be used, apart from the dimensions and the working pressures. The diameter (DN) of the pipes of the network depends on the size of water flow and the flow velocity that should be in the range of 1.4 m/s to 2.0 m/s (Table 17.1). The pipes working pressure (PN) must be around 6.0 Bars and in any case not less than 4.0 Bars. For the determination of the pipe diameter the flow velocity formula Q = AV is used in the form:

$$Q \text{ (l/h)} = V \text{ (m/s)} \times 2{,}826 \, di^2 \text{ mm}$$

where, **Q** the system flow, **V** the selected flow velocity (usually taken 1.7 m/s), **di** the pipe internal diameter. Based on the above, the recommended flow for various kinds and sizes of pipes are as follows:

TABLE 17.1 - Flow versus pipe diameter								
V = 1.7 m/s²	Galvanized steel threaded, light series			Rigid PVC, 6.0 bars			HDPE, 6.0 bars	
DN mm	2 in	2 ½ in	3 in	90	110	160	90	110
di mm	54	69	82	84.4	103.2	150.2	79	97
Flow/m³	14	23	32	34	51	108	30	45

COST

This system is a low-medium pressure with solid permanent installation. The initial cost per unit of area is lower as compared with the other closed pipes pressurized irrigation systems. The average cost per ha is around 850 US dollars. In the following example design the cost is US$900 per ha. Also the fuel consumption is lower than in any other improved irrigation system. Only the labor expenses are relatively high. This technique is classified as low-cost irrigation technology.

ADVANTAGES

- Low initial capital investment
- Availability of equipment
- Easy to operate and maintain
- High adaptability by the farmers
- Suitable for a wide range of crops
- Appropriate for all sizes of holdings and plot irregular shapes

DISADVANTAGES

- Requires skilled irrigator
- Low field application (delivery) efficiency
- In unleveled fields and sandy soils is not easily adapted
- Not applicable for small flows
- Best suited for medium-heavy soils

Despite the above mentioned advantages and disadvantages the main characteristic of this system is that this is the first step to be taken to facilitate the farmers to change from the traditional irrigation practices to the more advanced ones smoothly, safe and with the least expenses (Figure 17.4).

FIGURE 17.3 - The PDS Control Head and the Hydrant.

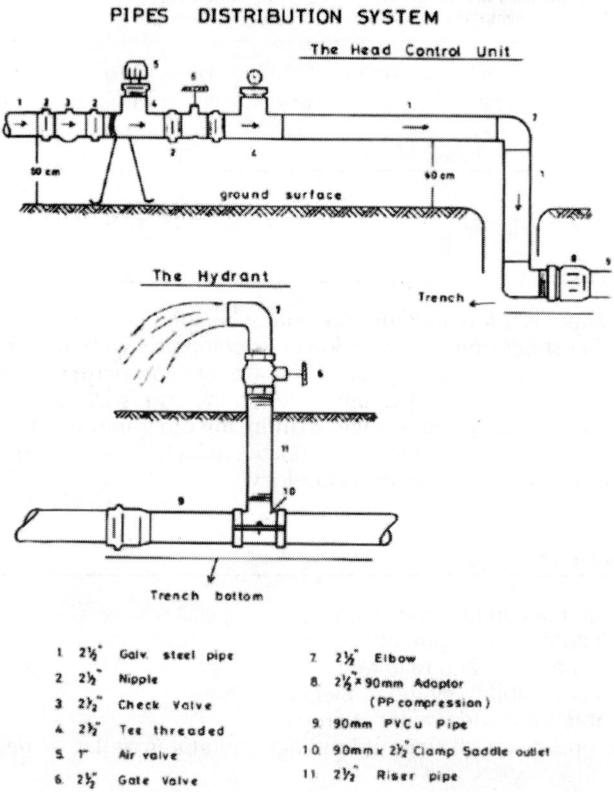

EXAMPLE DESIGN

Area and crop

The design area is 3.3 ha approximately, divided into six field-plots of the same rectangular shape and dimensions, 90 *x* 45 m. The crop can be citrus, cotton, vegetables, melons, potatoes, alfalfa, or any other seasonal or perennial one. The topography is smooth and there is a slope of around 0.4–0.6 percent from north to south and east to west (see map).

Soil and water

The soil is of medium texture and of good structure with moderate infiltration rate and internal drainage. **Sa** is around 150mm/m depth. The water is good quality with no toxicity or sodium hazards. It is a pumped from a nearby tube-well at a rate of 27 m³/hour (7.5 l/s) for 12 hours per day.

Water requirements and irrigation scheduling

The estimate of the irrigation needs and the schedule depends upon the climatic data and the kind of crop to be irrigated. However, the water availability that is 324 m³/day (27 m³/h *x* 12 hours) corresponds to a daily

FIGURE 17.4 - The Layout of the PDS.

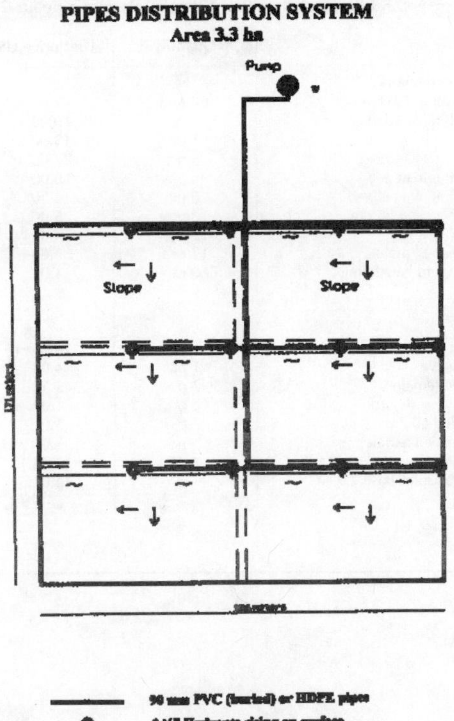

application of 9.8 mm in an area of 3.3 ha. This amount of water can meet the demands of any crop at peak.

The system layout and dynamic head

There is a conveyance pipeline, 100 m approx., from the pump to the farm gate. It is made of rigid PVC 90 mm, 6.0 Bars. The on-farm piping distribution network, of the same kind and size, is placed along the borderlines of the field plots. All pipes are buried. Take off hydrants 2 in are raising on surface, each one serving an area of 0.27 ha. The system's dynamic head for normal operation is the pipeline loss of head due to friction, plus the losses in the control head and the local minor losses, minus the difference in elevation, all included 0.9 Bars (Figure 17.5 and Table 17.2).

TABLE 17.2 - List of equipment needed for the Pipe Distribution System Installation (Bill of quantities). ***Area: 3.3 ha, System flow: 27 m^3/h, Dynamic head: 0.9 Bars. 0.9 Bars.***

Item	Description	Quantity	Unit price US$	Total price US$
	System distribution network			
1.	90 mm rigid PVC pipe, 6.0 bars	660 m	2.50	1650.00
2.	2 ½ in x 90 mm PP male adaptor	1 pc	10.00	10.00
3.	90 mm PP elbow	1 pc	15.00	15.00
4.	90 mm PP tee	5 pcs	22.00	110.00
5.	90 mm x 2 ½ in PP female tee	6 pcs	19.00	114.00
6.	90 mm x 2 ½ in PP female elbow	6 pcs	13.50	81.00
7.	2 ½ in threaded riser pipe 60 cm	12 pcs	5.00	60.00
8.	2 ½ in brass gate valve	12 pcs	14.00	168.00
9.	2 ½ in elbows threaded male	12 pcs	3.00	36.00
10.	Trench excavation and backfilling	660 m	1.00	660.00
	Sub-total			**2904.00**
	Head control			
11.	2 ½ in brass check valve	1 pc	16.00	16.00
12.	2 ½ in brass gate valve	1 pc	14.00	14.00
13.	2 ½ in tee threaded female	2 pcs	3.50	7.00
14.	2 ½ in nipple	2 pcs	1.00	2.00
15.	2 ½ in pipe threaded 60 cm	2 pcs	5.00	10.00
16.	2 ½ in elbows threaded female	1 pc	3.00	3.00
17.	1 in single air valve	1 pc	12.00	12.00
18.	Pressure gauge w/adaptor base 2 ½ in	1 pc	14.00	14.00
	Sub-total			**78.00**
	TOTAL COST			**2982.00**

CHAPTER 18: Low-cost hose irrigation

INTRODUCTION

In many countries the low-cost hose irrigation installations are popular among small and part-time farmers for the irrigation of many crops. This method of irrigation is an improvement on the traditional furrow-basin and furrow irrigation approaches. The water is applied to the basins and the furrows through ¾–1 ½ inch plastic hoses which are portable, 'hand-move', and can be extended in various directions. When one furrow or basin has been filled up with water, the hose is moved manually to the next one and so on.

A considerable engineering development has turned this practice from an old open surface method into a highly efficient closed pipe modern irrigation method. It is a localized method using a low pressure system, and a semi-permanent hand-move installation. It has been applied on a large scale and used extensively in many southern Mediterranean countries of the semi-arid zone in family managed smallholdings of about 1 ha. Properly designed hose basin systems for trees have also been successfully installed and operated on farms up to 20 ha (Figure 18.1).

FIGURE 18.1 - Hose basin irrigation in young fruit trees.

SYSTEM LAYOUT AND COMPONENTS

The layout of the system and the hydraulics of design and operation are almost the same as in the other closed pipe low pressure systems. On the main pipeline, there are hydrants with movable or permanently laid laterals running along the crop rows. Long hoses are connected on these lateral lines at a regular wide spacing to deliver water to each basin or furrow separately. Each hose covers many basins or furrows according to its length.

The system's piping network is also similar to the other low pressure irrigation systems. It can be either a complete installation with all component parts, as in the sprinkler and the micro-irrigation installations, or a simple one. A hose irrigation system usually consists only of a main pipeline of any kind, 50–90 mm (2–3 inches) uPVC, HDPE or layflat, 4.0–6.0 bars, which also serves as a manifold, with hydrants to which the laterals are connected. The laterals can be of any kind of 50–63 mm pipe but are usually LDPE, 4.0 bars. Smaller diameter long plastic hoses are connected on the laterals. Sometimes, the hoses can be fed directly from the source of the water, which can be a small reservoir at a higher level, a low capacity pump, or a tap. There is no need for filters, injectors or other accessories for a head control.

HOSES

The hoses are the well-known and widely available garden hoses. They are elasticized soft small diameter (¾–1 ½ inch) flexible PVC tubes with plain ends. Soft black 20–32 mm PE hoses (LDPE, 2.5–4.0 bars), are also used. The length of the hoses varies from 18 to 36 m and the water flow is 1.5–8.0 m^3/h. Thus, each hose can irrigate an area of approximately 600–2 100 m^2 respectively, covered with a number of small basins or furrows according to the cultivation. These sizes and lengths have been found to be the most convenient for farmers. The average flow characteristics for 24 m hoses with flow velocities up to 2.0 m/s are presented in Table 18.1.

TABLE 18.1 - Flow characteristics of 24 m hoses

Kind of hose	Nom. diameter	Average flow - m^3/h	Pressure losses - bars
Flexible PVC	¾ in	2.0	0.40
	1 in	3.6	0.30
	1 ¼ in	5.7	0.20
	1 ½ in	8.0	0.25
Soft polyethylene (LDPE)	20 mm	1.5	0.85
	25 mm	2.5	0.70
	32 mm	4.5	0.40

SYSTEM TYPES AND DESIGN CRITERIA

The types of hose irrigation systems refer to the characteristics of the water delivery hoses, their position in the field, the general operating procedure and the method of water distribution over the land (basin or furrow). In tree groves, each tree has a basin, whose shape and size is mainly determined by the age and the spacing of the trees. With close planting spacing, two to six trees can be in one larger rectangular basin along the row. With vegetables and other field crops, the land slope, type of soil, crop, water availability and farming practices determine the dimensions of the basins and furrows.

All types have movable water delivery hoses which are transferred or dragged from one spot to another. In this sense, there are four different types or variations of the system.

Conventional hose basin for trees

With a common tree spacing of 6 x 6 m, one 24 m-long hose can irrigate 36 tree basins in all directions in an area of about 1 300 m^2. The laterals are placed along the rows 36 m apart (every six rows) and the hoses are fitted on the laterals every six trees (36 m). Thus, the hose spacing is 36 x 36 m. With other planting spacings, the lateral and hose spacings differ, but not greatly, from the above (Figure 18.2). Flexible PVC garden hoses about 1 1/4 inch in diameter have proved to be the most suitable kind, as they can easily cross the field perpendicularly and diagonally without being damaged (cracked). The hoses are moved by hand from one basin to another.

FIGURE 18.2 - Hose furrow irrigation in vegetables.

Drag hose basin for trees

This type of system is an improvement on the conventional one as it is easier to design and operate. The water delivery hoses are 20–32 mm soft black LDPE pipes, 2.5 or 4.0 bars, connected to the laterals. Each hose can irrigate two or four rows of trees on both sides of the lateral line. The hose can be 20–40 m long, and the area covered from 900 to 1 800 m². It is called the drag hose system because at the beginning of each irrigation the hoses are extended to the distant end basins and then moved to the other basins by dragging them backwards (Figure 18.3).

FIGURE 18.3 - Schematic diagram for hose basin for trees.

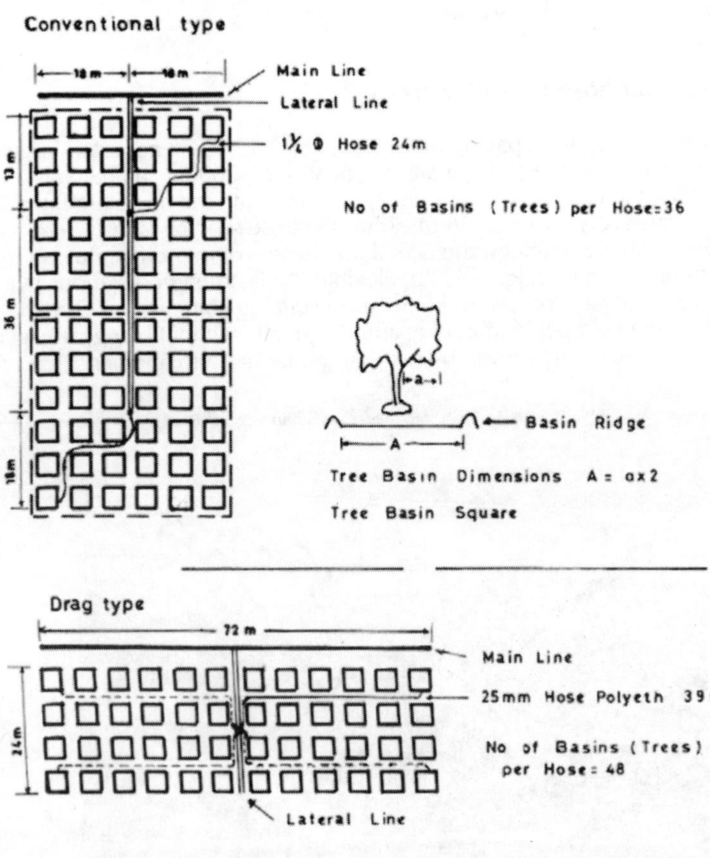

Hose basin for field crops

In this system, the hoses cannot cross the basins because they may damage the crop. The size of the small basins is usually 12 *x* 12 m, 6 *x* 12 m or 6 *x* 18 m. The laterals are placed at a closer spacing than for trees, in relation to the basins' dimensions and arrangement. The hoses can be of any kind, soft PVC or LDPE, in the appropriate lengths (18–24 m) and sizes (25 mm–1 1/4 inch). For example, with basins 6 m wide and 12 m long, the lateral lines are placed along the slope direction 24 m apart (every four basins). The 24 m hoses are connected to the laterals every three basin lengths (36 m), irrigating four basins upstream, two on each side, and eight basins downstream, four on either side, for a total of 12 basins, in an area of approximately 865 m^2. The hose spacing in this example is 24 *x* 36 m. However, it can vary as required. The hoses can be moved from one basin to another either by dragging them backwards or by carrying them.

Hose furrow for vegetables

Similar to the hose basin for vegetables, here the lateral lines are placed along the slope with the hoses connected at the head of the furrows. They are extended perpendicular to the lateral on either side delivering water to a number of furrows, as a drag system. The hoses are generally 25–32 mm soft black LDPE or 1 1/4 inch soft PVC garden hoses. The spacing of the hoses along the lateral is the same as the length of the furrows. The length of the furrows depends mainly on the type of soil, the slope, and the size of flow. With these systems, the furrows are usually short, 18–30 m long, 15 cm deep and about 1 m apart. The factors that influence the furrow layout are: farming practices; size and shape of the field; and irrigation application depth. The lower the depth of application and the size of the flow, the shorter the length; and the steeper the slope, the longer the furrow. In sandy soils, the furrow is shorter than in heavy clay soils. In medium texture soils, the following approximate relationship between slope and size of flow (Table 18.2) can be considered:

TABLE 18.2 - Slope and size of the flow

Slope %:	0.3	0.6	1.0	1.3	1.5
Size of flow m^3/h:	8.0	4.0	2.25	1.75	1.5

FIGURE 18.4 - Hose irrigation for vegetables & field crops. Hose positions and movements.

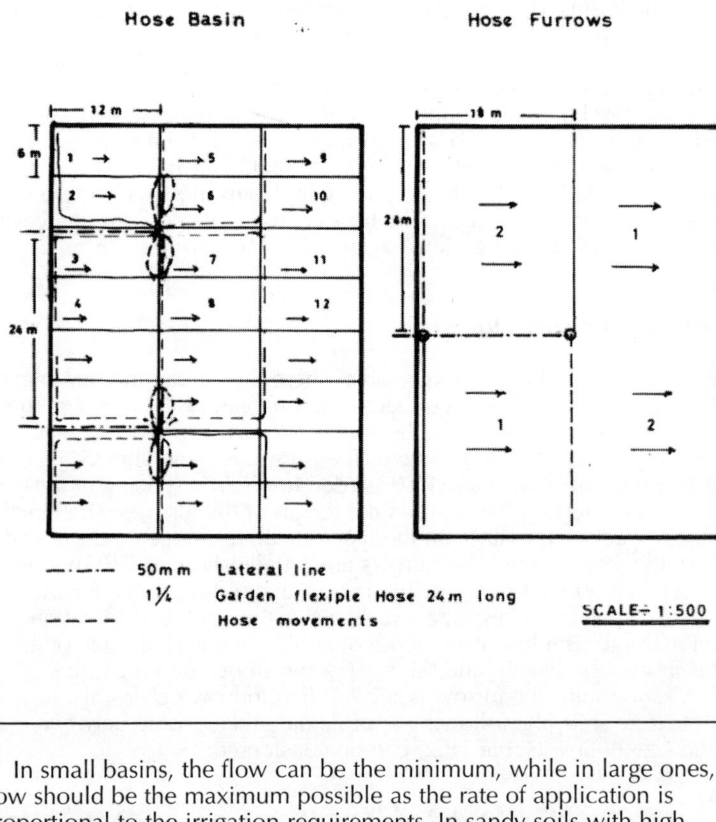

In small basins, the flow can be the minimum, while in large ones, the flow should be the maximum possible as the rate of application is proportional to the irrigation requirements. In sandy soils with high infiltration rates, the small discharge hoses can be moved from one place to another in the basin itself during the irrigation if necessary to ensure uniform distribution of water. It is common for farmers to subdivide the basins into smaller ones, or to construct short furrows within the basins in order to achieve ideal results. Several furrow layouts are applied (zigzag pattern, key pattern, etc.) in both types of systems. When the hose discharge is highly pressurized, some informal techniques are exercised on the spot by the farmer, such as the use of a tin vessel at the hose outlet, or a plastic bucket to avoid soil erosion and destruction of the ridges (Figure 18.4).

FIGURE 18.5 - Schematic diagram of short furrows.

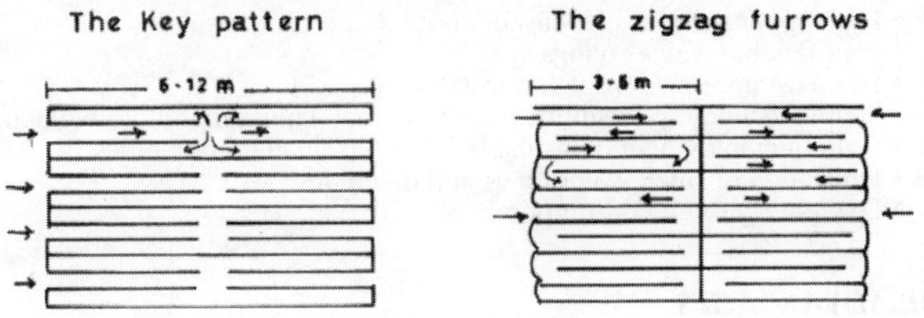

COST

Although hose irrigation systems are classed as semi-permanent installations, the water delivery hoses are the only movable component. However, the cost for a complete installation is very low compared with any other improved closed pipe technique. The average cost for all types of hose irrigation systems is about US$660/ha. Moreover, many years of study and observation have shown that the operating costs to the farmers, in terms of out-of-pocket money, are much lower than for any other system (Figure 18.5).

FIGURE 18.6 - Women irrigating young trees with hose basin system.

ADVANTAGES

- High application efficiencies of about 75 percent, resulting in considerable water savings.
- Low-cost improved irrigation installation.
- Simple technology easily managed by small children and old women.
- Gainful employment of available labour in small communities.
- Utilization of small water flows and quantities.
- Low energy (fuel) consumption.

DISADVANTAGES

- High labour requirements for system operation.

EXAMPLE DESIGNS – Hose basin with trees: conventional and drag types

Area and crop

The field dimensions (for design purposes) are taken as 140 *x* 72 m (about 1.0 ha) with mature trees planted in a spacing of 6 *x* 6 m. There are 24 rows with 12 trees in each row for a total of 288 trees (Figure 18.6).

Soil, water and climate

Medium texture soil of good structure, with good infiltration and internal drainage. Soil available moisture: 150 mm/m depth. The water is of good quality with no salinity or toxicity hazards; the source is a high-level reservoir. The peak irrigation demand is in July; the evaporation pan average readings are 7 mm/d.

Crop water requirements and irrigation scheduling

The pan reading of 7.0 mm/d multiplied by 0.66 (pan correction factor) gives an ETo of 4.65 mm/d. The crop factor kc is 0.65. Thus, ETc = 4.65 *x* 0.65 = 3.0 mm/d. The area shaded by canopy is 70 percent and for calculation purposes this is taken as 82 percent. Therefore, the daily water requirements are: 3.0 *x* 0.82 = 2.48 mm/d net. With a system application efficiency of 75 percent, the gross daily irrigation requirements are: 2.48 *x* 100 ÷ 75 = 3.3 mm (33 m^3). If irrigation takes place every ten days, the gross irrigation dosage is: 10 *x* 33 = 330 m^3.

The maximum permissible irrigation interval in July on a 50 percent moisture depletion for a trees root depth of 0.6 m is: 150 *x* 0.6 *x* 0.5 ÷ 3.0

FIGURE 18.7 - Hose irrigation for trees (example designs) (area=1.0 ha).

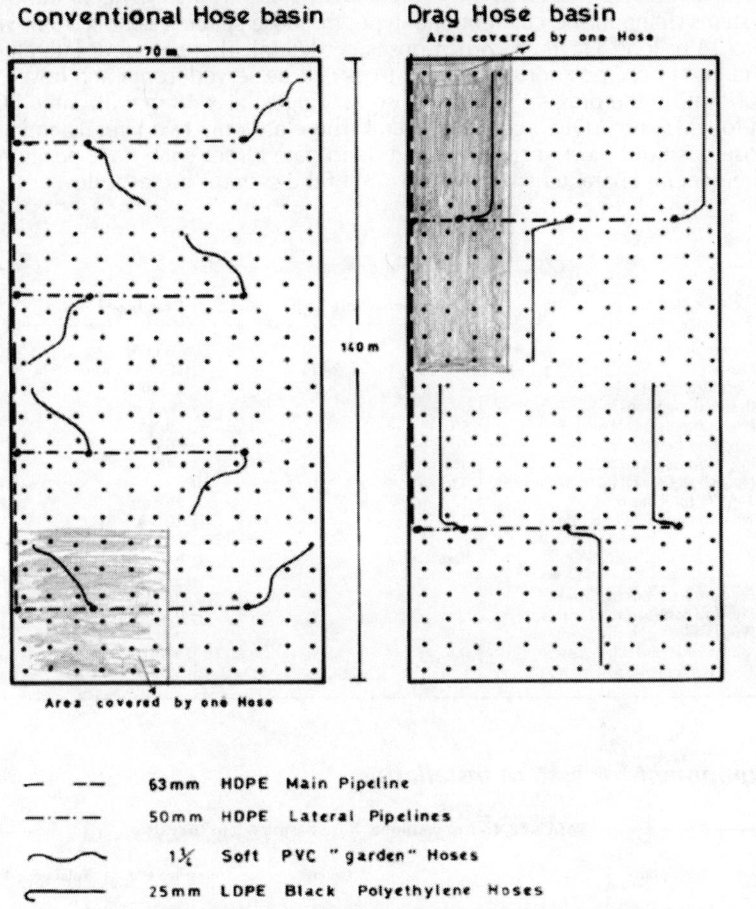

= 15 days. The irrigation frequency depends on many factors and in no case should exceed the maximum permissible irrigation interval.

Layout performance and hydraulics

In both the conventional and drag hose basin systems, a 63 mm HDPE or PVC main pipeline is placed along the border of the field with 2 in offtake hydrants, four for the conventional type and two for the drag type.

Laterals of 50 mm LDPE are laid perpendicular to the mains, four and two respectively, connected to the hydrants. The hose arrangements of the two systems differ. In the conventional type, the hose spacing is 36 x 36 m with two 24 m-long 1 1/4 inch garden hoses per lateral. There are four laterals and eight hose positions. Thirty-six trees can be served from each hose position. In the drag system, the hose spacing is 24 x 48 m with three 36 m-long 25 mm LDPE hoses per lateral. There are only two laterals and six hose positions. Forty-eight trees can be irrigated from each hose position. The general characteristics (Table 18.3) of the systems are as follows:

TABLE 18.3 - Conventional and drag hose basin

	Conventional hose basin	Drag hose basin
System flow	16 m^3/h	16 m^3/h
Hoses	Soft PVC, 1 1/4 in, 24 m	LDPE, 25 mm, 36 m
Hose discharge	4 m^3/h	2.7 m^3/h
Basins dimensions	5x 5 m	5x 5 m
No. of basins per hose	36	48
Irrigation frequency	10 days	10 days
Irrigation dosage	330 m^3	330 m^3
No. of hoses operating simultaneously	4 (double shift)	6
Time to fill a basin	17.4 min	26 min
Time to complete one irrigation	20.8 h	20.8 h
	bars	**bars**
Pressure losses in the hoses	0.2	0.7
Pressure losses in the laterals	0.3	0.4
Pressure losses in the main line	0.5	0.4
Minor local and other losses	0.5	0.5
Total dynamic head	1.5	2.0

Equipment for system installation

TABLE 18.4 - Conventional hose basin system (trees)

Item	Description	Quantity	Unit price US$	Total price US$
	System distribution network			
1.	63 mm HDPE pipe, 4.0 bars	125 m	1.80	225.00
2.	50 mm HDPE pipe, 4.0 bars	220 m	1.20	264.00
3.	63 mm PP compression end plug	1 pc	5.00	5.00
4.	50 mm PP compression end plug	4 pcs	3.00	12.00
5.	63 mm x 2 1/2 in PP compression adaptor	1 pc	5.00	5.00
6.	50 mm x 2 in PP compression adaptor	4 pcs	3.00	12.00
7.	63 mm x 2 in PP clamp saddle	4 pcs	1.30	5.20
8.	50 mm x 1 in PP clamp saddle	8 pcs	1.10	8.80
9.	2 in brass gate valve	4 pcs	8.00	32.00
10.	1 in brass gate valve	8 pcs	3.00	24.00
11.	2 in nipple	4 pcs	0.80	3.20
12.	1 in nipple	16 pcs	0.40	6.40
13.	1 1/4 in tap hose adaptor	4 pcs	0.70	28.00
14.	1 1/4 in soft PVC garden hose, 24 m long	4 pcs	30.00	120.00
	TOTAL COST			**750.60**

TABLE 18.5 - Drag hose basin installation (trees)

Item	Description	Quantity	Unit price US$	Total price US$
	System distribution network			
1.	63 mm HDPE pipe, 4.0 bars	105 m	1.40	147.00
2.	50 mm HDPE pipe, 4.0 bars	120 m	0.80	96.00
3.	63 mm PP compression end plug	1 pc	5.00	5.00
4.	50 mm PP compression end plug	1 pc	3.00	3.00
5.	63 mm *x* 2 ½ in PP compression adaptor	1 pc	5.00	5.00
6.	50 mm *x* 2 in PP compression adaptor	2 pcs	3.00	6.00
7.	63 mm *x* 2 in PP clamp saddle	2 pcs	1.30	2.60
8.	50 mm *x* 1 in PP clamp saddle	6 pcs	1.10	6.60
9.	2 in brass gate valve	2 pcs	8.00	16.00
10.	1 in brass gate valve	6 pcs	3.00	18.00
11.	2 in nipple	2 pcs	0.80	1.60
12.	1 in nipple	6 pcs	0.40	2.40
13.	1 in *x* 25-mm PP compression elbow	6 pcs	1.20	7.20
14.	25-mm soft LDPE hose, 4.0 bars, 36 m	6 pcs	14.40	86.40
	TOTAL COST			**402.80**

FIGURE 18.8 - Hose irrigation for early vegetables in low tunnel.

HOSE BASIN IRRIGATION FOR MAIZE AND HOSE FURROW FOR TOMATOES

Area and crops

Two plots of the same dimensions 108 *x* 96 m, 1.0 ha area each, are planted with maize and tomatoes respectively in mid-April. The maize plot is arranged in small basins, 6 *x* 12 m, for a total of 144 basins. The tomato plot is planted in short furrows 18 m long (Figure 18.8).

Soil, water and climate

Medium texture soil of good structure, with good infiltration and internal drainage. Soil available moisture: 150 mm/m depth. The water is of good quality with no salinity or toxicity hazards; the source is a high-level reservoir. The peak irrigation demand is in July; the evaporation pan average readings are 7 mm/d.

Crop water requirements and irrigation scheduling

The pan reading of 7 mm/d multiplied by 0.66 (pan correction factor) gives an ETo of 4.65 mm/d. The crop factor for maize at harvest time in July is taken as 1.0 and for late season tomatoes as 0.85. Thus, ETc maize: 4.65 *x* 1 = 4.65 mm/d, and ETc tomatoes: 4.65 *x* 0.85 = 3.95 mm/d. The system's application efficiency is 75 percent. Therefore, the gross daily irrigation requirements at peak are: 4.65 ÷ 0.75 = 6.2 mm (62 m^3) for maize; and 3.95 ÷ 0.75 = 5.26 mm (52.6 m^3) for tomatoes. At peak demand, the irrigation frequency for maize is every seven days, while for tomatoes it is every other day. Thus, the irrigation dosage would be: 7 *x* 62 = 434 m^3 for maize; and 2 *x* 52.6 = 105 m^3 for tomatoes. The available system flow is 16 m^3/h.

Layout, performance and hydraulics

In both systems, the pipe network is almost the same, i.e. 63 mm HDPE pipes for the main line, 50 mm HDPE pipes for the lateral lines, and 1 ¼ inch flexible PVC garden hoses. In the basin system, the hose spacing is 36 m along the laterals and 12 m between the laterals. There are four laterals with three hoses each for a total of 12 hose positions. Each hose position can serve 12 basins.

In the short furrow system, the hose spacing is 18 m along the lateral and 24 m between the laterals. There are only two laterals with six hoses each, for a total of 12 hose positions irrigating the furrows downstream on either lateral side. The general characteristics (Table 18.6) of the systems are:

FIGURE 18.9 - Hose irrigation for vegetables and field crops.

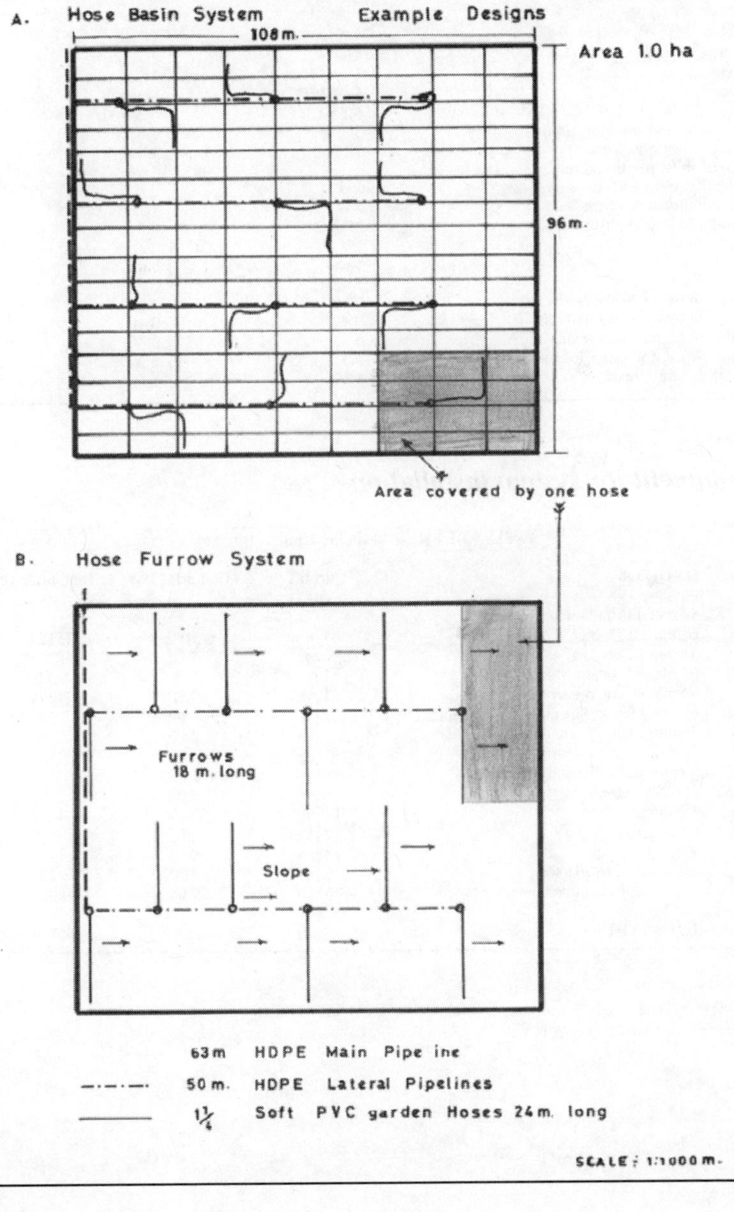

TABLE 18.6 - Hose basin and hose furrow

	Hose basin	Hose furrow
Area	1.0 ha	1.0 ha
Crop	Maize	Tomatoes
System flow	16 m^3/h	16 m^3/h
Hose type	Soft PVC, 1 ¼ in, 24 m	Soft PVC, 1 ¼ in, 24 m
Hose discharge	5.3 m^3/h	5.3 m^3/h
No. of hoses operating simultaneously	3	3
No. of hose positions	12	12
No. of shifts per irrigation	4	4
Irrigation frequency at peak	7 d	2 d
Irrigation dosage (gross)	434 m^3	105 m^3
Time to complete irrigation	27 h	6.5 h
	bars	**bars**
Pressure losses in the hoses	0.20	0.20
Pressure losses in the laterals	0.40	0.40
Pressure losses in the main line	0.45	0.40
Minor local and other losses	0.55	0.50
Total dynamic head	1.60	1.50

Equipment for system installation

TABLE 18.7 - Hose basin installation (maize)

Item	Description	Quantity	Unit price US$	Total price US$
	System distribution network			
1.	63 mm HDPE pipe, 4.0 bars	84 m	1.80	151.20
2.	50 mm HDPE pipe, 4.0 bars	336 m	1.20	403.20
3.	63 mm PP compression end plug	1 pc	5.00	5.00
4.	50 mm PP compression end plug	4 pcs	3.00	12.00
5.	63 mm x 2 ½ in PP compression adaptor	1 pc	5.00	5.00
6.	50 mm x 2 in PP compression adaptor	4 pcs	3.00	12.00
7.	63 mm x 2 in PP clamp saddle	4 pcs	1.30	5.20
8.	50 mm x 1 in PP clamp saddle	12 pcs	1.10	13.20
9.	2 in brass gate valve	4 pcs	8.00	32.00
10.	1 in brass gate valve	12 pcs	3.00	36.00
11.	2 in nipple	4 pcs	0.80	3.20
12.	1 in nipple	12 pcs	0.40	4.80
13.	1 ¼ in tap hose adaptor	6 pcs	0.70	4.20
14.	1 ½ in soft PVC garden hose, 24 m long	6 pcs	30.00	180.00
	TOTAL COST			**867.00**

TABLE 18.8 - Hose furrow installation (tomatoes)

Item	Description	Quantity	Unit price US$	Total price US$
	System distribution network			
1.	63 mm HDPE pipe, 4.0 bars	72 m	1.80	129.60
2.	50 mm HDPE pipe, 4.0 bars	180 m	1.20	216.00
3.	63 mm PP compression end plug	1 pc	5.00	5.00
4.	50 mm PP compression end plug	2 pcs	3.00	6.00
5.	63 mm x 2 ½ in PP compression adaptor	1 pc	5.00	5.00
6.	50 mm x 2 in PP compression adaptor	2 pcs	3.00	6.00
7.	63 mm x 2 in PP clamp saddle	2 pcs	1.30	2.60
8.	50 mm x 1 in PP clamp saddle	12 pcs	1.10	13.20
9.	2 in brass gate valve	2 pcs	8.00	16.00
10.	1 in brass gate valve	12 pcs	3.00	36.00
11.	2 in nipple	2 pcs	0.80	1.60
12.	1 in nipple	12 pcs	0.40	4.80
13.	1 ¼ in tap hose adaptor	6 pcs	0.70	4.20
14.	1 ½ in soft PVC garden hose, 24 m long	6 pcs	30.00	180.00
	TOTAL COST			**626.00**

FIGURE 18.10 - Hose irrigation jointing techniques.

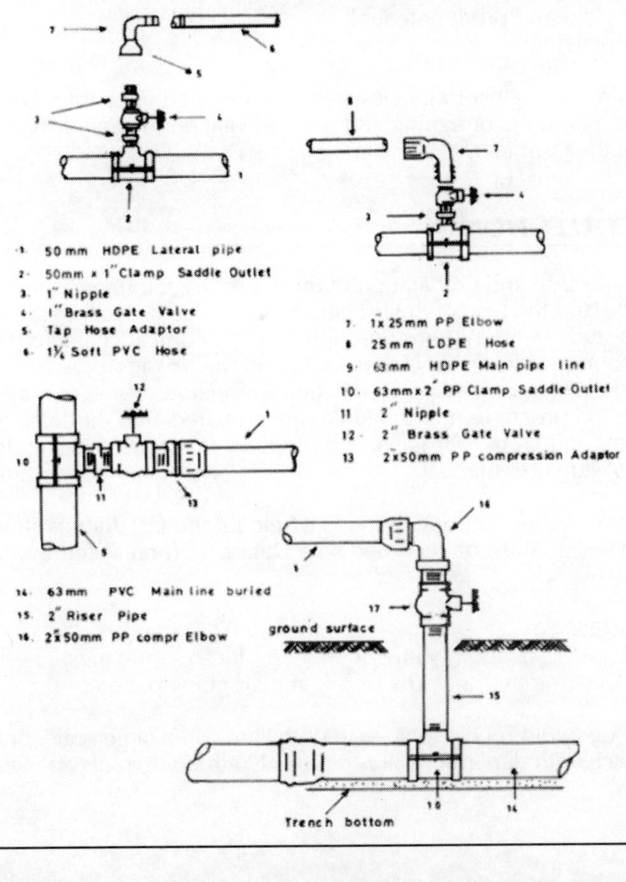

CHAPTER 19: An outline for engineering investigation for a pressurized irrigation system

INTRODUCTION

The choice of the system of irrigation sometimes is predetermined by specific limiting factors, which leave no alternatives. In other cases, where more than one system is theoretically possible, the final choice is made on the basis of sound criteria, as:

- The suitability/adaptability under the prevailing conditions,
- The cost,
- The efficiency (water savings),
- The layout flexibility,
- The yield and profit potential,
- The sustainability

A minimum engineering investigation is required to enable the successful planning, designing and implementation of every irrigation system at the farm level.

DATA COLLECTION

The collection and preparation of the necessary information are related to the kind and the type of the irrigation system and its techniques. A thorough study of the systems description and technical characteristics is of major importance for the selection criteria of the irrigation systems. Detail plans and designs with maps, installation instructions, jointing drawings and irrigation programs and schedules are prepared after the selection of the systems and crops to irrigate. The suppliers of the systems should always provide user manual.

The data needed for each individual field for the installation of the irrigation system must be recorded on a Datasheet form as follows:

Farm Datasheet

a) *Farm identification:*Name of the farm, location, ownership (private or Government), size (ha or m^2) and cropping pattern.

b) *Topography:* Topographic map of the area on a large scale, or drawing sketch with dimensions illustrating – North point, plots arrangement and dimensions, location of water source, farm roads, premises etc. – Contour lines, or elevation points and direction of slopes.

c) *Crops:* Kind, area and location of each crop on the map - Age if perennials. Cropping pattern for annuals – planting spacing along and between the rows – direction of rows – Height of plants - Growing season/irrigation period - kc factors (crop coefficient).

d) *Soil:* Type and physical characteristics, e.g. sandy, sandy, sandy loam, silt loam, clay loam – Permeability, internal drainage – water holding capacity – depth of top soil – existence of hardpan – Potential problems of salinity, toxicity, alkalinity.

e) *Climate and altitude:*Height of farm above sea level – Rainfall (monthly totals averaged over the last five years) and effective rainfall – Temperature (monthly average maximum) – Relative humidity Winds prevailing (direction and velocity) - ETo values

f) *Water:* Source of water supply and type (deep borehole, spring, river, other) - Location (distance apart and difference in elevation) - Flow available (m^3/h or l/s) and quantity per day when at lowest levels – Quality, physical (foreign suspended particles content such as sand, silt, impurities, algae etc), chemical (complete ionic analysis plus boron and nitrates) and biological if treated wastewater - Depth of borehole, static water level, draw down and safe capacity – Type, capacity and output of existing pumping unit available.

g) *Existing conditions:*Existing water conveyance network - Present irrigation method (frequency of applications, operating hours and quantities applied) – Equipment available.

h) *Labour availability* and average working hours in the fields – Maximum recommended daily operating hours of the improved irrigation installations.

i) *Remarks and recommendations:* Any other information of particular importance, Remarks and recommendations.

As it is concerned with the water availability it must be noted that in cases where the source of water is far from the command field to be irrigated, then a conveyance pipeline should be installed from the source to the field. The distance and the difference in elevation in no one case should affect the pressure needed for the normal operation of the systems. A Booster pump can either be placed at the beginning of the conveyance pipeline or at the beginning of the system. Arrangements should be planned according to the site conditions.

SELECTION CRITERIA AND PARAMETERS FOR VARIOUS SYSTEMS

The criteria and data collected are examined and evaluated in accordance with the various irrigation systems technical characteristics and performances. Here below are the main parameters considered for the selection of a kind of system. Additional parameters and factors are examined too, such as, easy accessibility, protected area, organized farm, labour availability, operation and maintenance facilities etc.

Note:
Description and characteristics of the systems are given in the related chapters.

The center pivot (CP) irrigation systems

Kind of crops

Nearly all crops can be grown under CP irrigation. The field crops mostly recommended are the cereals, agro-industrial, leafy vegetables and the forage crops.

Area, size and shape

The area should be a plain agricultural field of a relatively large size 15–100 ha. Pivot systems can be tow able and moved to a next position nearby and so on. This practice is usually applied in cereals for supplementary irrigation during drought periods.

Topography

These CP irrigation systems can operate on uneven ground; however, level lands are recommended and uniform sloping fields with slopes up to 3 percent. Undulating topography may produce a lot of difficulties especially where runoffs occur.

Soil

The soil should be of medium texture with high infiltration rate >15 mm/hour good internal drainage and water holding capacity.

Water availability

The source of water can be a tube-well, a river, a small water tank. But the CP systems, like all circular CP systems, will always be fed from a hydrant placed at the centre of the irrigable area near the pivot. So a buried water conveyance pipeline should be installed from the source of the water to the pivot. At the end of the pipeline, in the middle of the area near the pivot the hydrant should be installed to deliver irrigation water at pressure of about 3.0 Bars. The system inlet will be connected to the hydrant through a quick coupling flexible hose. For every CP position a hydrant is needed.

Water quality

The water should be clean and free from suspended solids and other impurities, of normal pH 6.5 to 8.4, with no salinity hazard, sodium hazard and toxicity problems caused by bicarbonates, nitrates or boron. TDS should not exceed 1 500 mg/l (ppm), SAR < 12, RSC < 1.25 meq/l, Boron content < 0.7 mg/l, Chlorides <200 mg/l, Nitrates (NO_3) < 100 mg/l and low content Bicarbonates (HCO_3).

Fuel requirements

The CP systems are equipped with generators for driving the towers and booster pumps, both diesel engine driven. Arrangements should be made so that the fuel tanks to be connected with additional bigger tanks placed nearby for long uninterrupted operation of the CP systems.

Traveller irrigation machines spray boom carts

Kind of crops

The field crops to be grown, among others, under Spray boom irrigation are the same as with the Center pivots such as wheat, barley, chickpeas, lentils, potatoes, industrial crops soybeans, maize, sunflower, leafy vegetables, water melons, alfalfa, perennial etc. Their largest application is for supplementary irrigation of cereals (wheat and barley) during the winter months. The height of the plants is considered for the boom "ground clearance".

Area, size and shape

The area should be a plain agricultural field of at least 1.8 ha. The system is tow able it can be towed and moved to a next position nearby and so on. This practice is usually applied in cereals for supplementary irrigation during drought periods.

Topography

The Spray boom irrigation systems can operate on uneven ground, however, level lands are recommended and uniform sloping fields with slopes up to 1 percent. Undulating topography may produce a lot of difficulties especially where runoffs occur.

Soil

The soil should be of medium texture with high infiltration rate >15 mm/h good internal drainage and water holding capacity.

Water availability

The source of water can be a tube-well, a river, a small water tank. The system can be fed with water from hydrants placed at various points on the farm plot boundaries. The system inlet is connected to the hydrant or the pump outlet through a quick coupling flexible hose. For every Spray boom position a hydrant is needed. The water source should be as near as possible too the field. The water pressure should be from 3.5 to 5.0 Bars.

Water quality

The water should be clean and free from suspended solids and other impurities, of normal pH 6.5 to 8.4, with no salinity hazard, sodium hazard and toxicity problems caused by bicarbonates, nitrates or boron. TDS should not exceed, if possible, 1 500 mg/l (ppm), SAR < 12, RSC < 1.25 meq/l, Boron content < 0.7 mg/l, Chlorides <200 mg/l, Nitrates (NO_3) < 100 mg/l and low content Bicarbonates (HCO_3).

The drip irrigation system

Kind of crops

The field crops, which can be grown, among others, under Drip irrigation techniques are all crops planted in rows and mainly vegetables in winter and summer time and water melons. Deciduous trees give excellent results with drip irrigation.

Area, size and shape

The area can be any farm planted with field crops in rows of any length from 40 to 150 meters length located in the mountains or in the plains. The size of plots can be from 0.2 to 1.0 ha. The shape should be of normal rectangular or square shape.

Topography

The drip irrigation systems whether with normal dripper emitters or with the pressure compensated ones that can operate on uneven ground, level lands are recommended and uniform sloping fields with slopes up to 3 percent.

Type of soil

The soil can be of any texture, preferably medium and/or fine and with infiltration rate < 20 mm/h. Very light sandy soils with high permeability are not recommended.

Water availability

The source of water can be a tube-well, a river, a small water tank. In cases of a booster pump, this should be placed at the beginning of the system before the Head control unit. The system operating pressure is around 3.0 Bars.

Water quality

The water should be, as clean as possible although there must be a complete filtration system. Chemically it must be of normal pH 6.5–8.4, with low to medium salinity, low sodium hazard and toxicity problems caused by bicarbonates, nitrates or boron. TDS can be from 500 up to 2 000 mg/l (ppm), SAR < 12, RSC < 1.25 meq/l, boron content< 0.9 mg/l.

Mini-sprinklers irrigation systems (for fruit trees)

Kind of crops

Any fruit trees, which can be grown in the area.

Area, size and shape

The area can be any agricultural field located in the mountains or in the plains and it is or will be planted with fruit trees in rows of a maximum length of 80 to 90 meters even ground. The size of separate plots can be from 0.5 to 1.0 ha. The shape should be of normal rectangular or square shape.

Topography

The Mini-sprinkler irrigation systems normally operate on smooth even ground, so level lands are recommended and uniform sloping fields with slopes ranging from 0.25 percent to 5 percent. In sloppy lands the length of the rows and the Mini-sprinkler lines vary accordingly.

Type of soil

The soil can be of any texture, preferably medium and/or fine, but with infiltration rate > 6 mm/h. Very light sandy soils with high permeability are suitable too.

Water availability

The source of water can be a tube-well, a river, a small water tank. The system operating pressure should be 2.5–3.0 bars.

Water quality

The water should be, as clean as possible although there is a disc-filter. Chemically it must be of normal pH 6.5–8.4, with low to medium salinity, low sodium hazard and toxicity problems caused by bicarbonates, nitrates and especially chlorides and boron. TDS can be from 500 up to 1 500 mg/l (ppm), SAR < 12, RSC < 1.25 meq/l, Cl < 12 meq/l, boron content< 0.7 mg/l.

Pipe distribution system

Kind of crops

The field crops, which can be grown, among others, under the Pipe Distribution irrigation techniques are nearly all winter and summer crops annual and perennial, i.e. vegetables, cereals, melons, forage crops and fruit trees.

The system consists of a water conveyance and distribution pipe network for surface irrigation methods in-plot. It is actually the replacement of the open channels with a properly designed closed piping network to convey and distribute the irrigation water from the source to the field plots without any losses.

Area, size and shape

The area can be any agricultural farm, planted with field crops irrigated with furrows, borders, basins or any other method of water application, located preferably in more or less plain areas. The size of the field plots can be from 0.1 to 1.0 ha, planted with one or more cultivations. Normally rectangular or square shapes are recommended.

Topography

The Pipe Distribution irrigation systems network can be installed and operate on uneven ground, however, the systems hydrants should be placed at the highest points of each plot. The method of application is surface, so level lands and uniform sloping fields with regular slopes of 0.1 to 0.25 percent are recommended.

Type of soil

The soil can be of any texture, but preferably of medium and with infiltration rate < 20 mm/h. Very light sandy soils with very high permeability are not recommended.

Water availability

The source of water can be a tube-well, a river, a small water tank. The pressure of the system is 1.0 to 2.0 Bars.

Water quality

The water should be, as clean as possible. Chemically it must be of normal pH 6.5–8.4, with low to medium salinity, low sodium hazard and toxicity problems caused by bicarbonates, nitrates or boron. TDS can be from 500 up to 2 500 mg/l (ppm), SAR < 12, RSC < 1.25 meq/l, boron content< 0.75 mg/l.

Hose-move sprinkler irrigation system

Kind of crops

Alfalfa and other forage crops, maize, sunflower, and other dense planted crops. This system can be installed successfully for nurseries.

Area, size and shape

The area can be any agricultural levelled farm located in the mountains or in the plains planted with agricultural crops. The plots size can be from 0.5 to 1.0 ha. The shape should be of normal rectangular or square shape.

Topography

The Hose-move sprinkler irrigation systems normally operate on smooth even ground, so level lands are recommended and uniform sloping fields with slopes ranging from 0.25 percent to 0.2 percent.

Type of soil

The soil can be of any texture, preferably medium and/or fine, but with infiltration rate > 8 mm/h and good internal drainage. Very light sandy soils with high permeability are suitable too.

Wind Conditions

Wind directions and velocities must be recorded and classified accordingly, (0–0.7 m/s nil wind, 0.7–2.5 m/s light, 2.5–3.5 m/s moderate to strong, and > 3.5 m/s very strong). Sprinkling is not recommended under strong wind conditions.

Water availability

The source of water can be a tube-well, a river, a small water tank. The designed flow of the system at 3.5 bars pressure.

Water quality

The water should be of good quality suitable for irrigation purposes. Chemically it must be of normal pH 6.5–8.4, with low to medium salinity, low sodium hazard and toxicity problems caused by chlorides, bicarbonates and nitrates. TDS can be from 500 up to 2 000 mg/l (ppm), SAR < 12, RSC < 1.25 meq/l, boron content< 1.0 mg/l.

Low-cost (family kid) drip irrigation systems

Kind of crops

The field crops, which can be grown, among others, under Drip irrigation techniques in N. Iraq and their growing season are mainly vegetables in winter and summer time and water melons.

Area, size and shape

The area can be any agricultural farm, planted with field crops in rows of short lengths from 12 to 24 meters located in the rural areas on the mountains or in the plains. The size can be from 250 m to 1 000 m. The shape should be of normal rectangular or square shape.

Topography

These drip irrigation systems operate at very low pressures level lands are recommended and uniform sloping fields with slopes < 0.5 percent.

Type of soil

The soil can be of any texture, preferably medium and/or fine and with infiltration rate < 20 mm/h. Very light sandy soils with high permeability are not recommended.

Water availability

The source of water can be a small well, a tub, a garden hose or anything that can fill the system's water tank regularly. The designed flow of the Family drip irrigation systems is around 1.1 m/h at 1.5 m head.

Water quality

The water should be, as clean as possible although there is a complete filtration system. Chemically it must be of normal pH 6.5–8.4, with low to medium salinity, low sodium hazard and toxicity problems caused by bicarbonates, nitrates or boron. TDS can be from 500 up to 2 000 mg/l (ppm), SAR < 12, RSC < 1.25 meq/l, boron content< 0.9 mg/l.

CHAPTER 20: Operation and maintenance

INTRODUCTION

The efficient operation of an irrigation system depends mainly on the ability of the farmer to make the best use of it. For every system, depending on the kind and the type of the installation and the way the water is delivered to the farm, there are several steps to be taken and factors to consider in order to ensure the efficient operation and performance of the installation. Sometimes, the irrigation installation fails to give full satisfaction because of poor design, faulty installation, or equipment that does not conform to specification. However, the way both the irrigation system as a whole and its component parts are operated and maintained will determine the success or failure of any properly designed and installed system.

The O&M of the irrigation system is also the key factor for good irrigation management. Farmers need a sound knowledge of the O&M procedures of their installation. This knowledge should be acquired from complete information, demonstration and written instructions from the designers and the suppliers.

OPERATION

When and how long to irrigate

The application of the exact amount of water required by the crops at the right time is the main achievement of the irrigation installation. Farmers normally understand matters concerning the main elements of irrigation programming, such as water discharge and rate, operating hours and irrigation frequency, and they can follow instructions. Properly installed, operated and maintained irrigation networks enable farmers to exercise absolute control over water use at farm level. Thus, it is easy for them to apply irrigation schedules based on crop, soil, weather, and water availability and quality factors.

Starting and stopping the system

Starting and shutting down the pressurized irrigation installation needs to be done very carefully in order to prevent surges and water hammer and to avoid air pockets in the pipelines.

The opening and closing of the valves at the head of the system, the main and submain pipelines, should always be done slowly.

Where there is a pump, engine or motor driven, the supplier's instructions should be followed. Priming the pump, filling the pipes, adjusting the speed and lubricating the pumping equipment are matters of major importance. Manufacturers provide detailed instructions in their literature for starting and operating each pumping unit.

System performance

Frequent observations and checks should be carried out during the irrigation season to ensure the proper functioning and good performance of the system. This involves a number of procedures for simple evaluations based on measurements taken under field conditions.

The equipment needed for this task is as follows:

- a map/sketch of the irrigated area showing the locations of all the system's component parts and the various plots;
- a portable pressures gauge (0–6.0 bars) with a special adaptor and pivot tube adjustment;
- a stopwatch (chronometer);
- a measuring tape, approximately 20 m;
- graduated vessels, capacity 1–5 litres;
- a soil auger, shove or probe;
- a notebook for recording data.

In most closed pipe pressure systems there are a number of factors that should be evaluated to determine the level of operation and that can be readjusted where not satisfactory.

Operating pressures

With the system in operation, pressure measurements are taken at various points on the piping network, preferably at the beginning and at the far end of the main and the submain pipelines. The operating pressures of the first and last emitters on a number of laterals are also measured. All pressures should be within the designed range. The difference in the emitter pressure should not differ from the recommended average pressure by more than 20 percent on level ground. Any change should be investigated immediately.

Flow rates and water discharge

The flow rates (discharge) of the same emitters whose operating pressures are measured are also determined. This is done by recording the time required to fill up a graduated vessel with water. The figures should be in accordance with the supplier's specifications and the difference between them should be less than 10 percent. The system's rate of discharge is the sum of the emitters' average flow rates.

Uniformity of application and depth of wetting

This may be checked by probing the soil at various locations using a probe, shovel or soil auger. The examination can be made 12–24 h after irrigation depending on the type of soil. Water should penetrate a few centimetres below the root depth. Areas taking less or more water can be easily identified for further investigation.

Visual observations for evaluation purposes of any type should be avoided as they might lead to misjudgements. In addition to the above simple evaluations, the following checks, on-site modifications, re-arrangements and preventive maintenance are necessary:

- Check and repair any leakage in piping or through valves.
- Position the sprinklers vertically to the ground and check spacing.
- Replace or rehabilitate clogged emitters.
- Flush the system network at least three times during the irrigation season when water originates from underground. With reservoir water, flush once every fourth irrigation. An approximate flushing time of 2–3 min for each line will prevent sedimentation on the inner pipes walls.
- Clean the filters of the system thoroughly before every irrigation. During operation, check for the minimum difference in pressure between the inlet and the outlet of the main filter.
- Check the air and check valves periodically for proper functioning.
- Inspect plastic equipment, valves and devices for cracks and other physical damage.
- Flush fertilizer injectors (pump and tank) after each use. Inspect hoses and valves.
- Conduct systematic checks to spot malfunctioning equipment affected by physical deterioration and other possible damage by machinery, animals, etc.
- Make frequent visual checks of the system to ensure that it is in good condition and operating efficiently.

Pump plant

Preventive maintenance of the pumping system is essential during the irrigation season. Equipment manuals contain trouble-shooting chapters which are useful for solving common problems associated with the normal operation of the pumping unit. The following checks and inspections are recommended for most engine or electric motor driven pumps:

- noise;
- vibration;
- leakage;
- temperatures of bearings and windings;
- fuel/power consumption;
- capacity and output (water discharge and dynamic head);
- ventilation screens, clean where necessary;
- oil pressure;
- oil, lubrication, change where necessary.

MAINTENANCE

The long-term operation of the irrigation installation depends upon simple maintenance carried out by the farmer. The periodic servicing of pumping plants and the repair of special devices (filters, injector, etc.) is carried out by trained maintenance and repair personnel.

Maintenance is carried out during a period of non-use to prepare the system: a) for the off-season shut-down; and b) for use before the next season. All equipment requires a certain amount of care in handling for storage and maintenance. For every installation there is a procedure which concerns various aspects of the distribution network and the pumping unit.

System network

The procedure for the network is as follows:

- Flush mains, submains, manifolds and laterals.
- Inspect for possible damage to the network and repair it.
- Open fully and drain completely all valves.
- Remove dirt, corrosion and other foreign material from the component parts.
- Check emitters for possible clogging, damage, wear and signs of deterioration, and replace where necessary.
- Store all emitters in a dry clean place on shelves away from fertilizers, chemicals, oil, grease and lubricants.
- Examine the condition of air and check valves.
- Flush and drain filtration and fertilizer injection equipment.

- Clean all filter elements.
- Check condition of gaskets and seals; remove, clean and store in a dry place.
- Retrieve all portable plastic tubes by rolling them up in coils; store properly.
- Inspect all portable metal pipes for any kind of damage and consult suppliers for repair; store properly away from power lines and wiring.
- Drain completely all pipes left in the open.

Pump plant

Pump plants usually consist of a centrifugal pump of some type and the power unit (electric motor or internal combustion engine). Maintenance instructions are available from manufacturers, pump users associations and other technical organizations. Special care should be taken to protect engines from moisture that can accumulate inside the machines and cause serious damage.

Below is a list of checks, inspections and steps to be taken for the preparation of the pumping plant a) for the off-season period and b) for use before the next season:

Maintenance for the off-season period

Centrifugal pumps

- Drain all the water from pump and connecting pipelines.
- Where possible, remove suction lines and store them.
- Cover shaft and any exposed metal and all oil or grease lubricated bearings with protective lubricant.
- Loosen 'v' belt or flat belt drive and insert piece of greaseproof paper between belts and pulley.
- Loosen packing gland.
- Clean debris and any other material from impeller and volute.

Internal combustion engines

- Run engine to thoroughly warm up oil in the crankcase; stop engine and drain crankcase oil; replace drain plug and refill crankcase with high-grade engine oil; start engine and run slowly for two minutes to complete oil distribution on all surfaces.
- Stop engine; remove all spark plugs; pour 60 ml of engine oil into each spark plug hole; with ignition switch off, crank engine for several revolutions to distribute oil over the cylinder walls and valve mechanism; replace spark plugs.

- Drain oil from crankcase; drain cooling system and close drain cocks; drain all fuel from tank, lines and carburettor bowl; replace all plugs and close drain cock.
- Lubricate all accessories and seal all openings airtight, including air cleaner inlet, exhaust outlet, and crankcase breather tube, with weatherproof masking tape.
- Check oil filler cap, gas tank and radiator cap.
- Spray all accessories and electrical equipment with suitable insulating compound.
- Insert a strip of greaseproof paper under the 'v' belt pulley.
- Remove battery and store fully charged.
- Where the engine is in the open, cover with waterproof material.

Electric motors

- Ensure that all bearings are well lubricated.
- Cover motor to protect against rodents, insects and dust, but provide ventilation.
- Lock control box in 'off' position and cover with a canvas where exposed in the open to protect against moisture and dust.

Preparation for use before the next period

Centrifugal pumps

- Where there is a trash screen, clean and install it properly.
- Ensure foot valve on suction line of horizontal centrifugal pumps operates properly.
- Install suction line of horizontal pumps and/or vertical turbine pumps and/or check they are adequately submerged; check impeller adjustment of deep-well vertical turbine pumps.
- Clean all passages for liquid.
- Tighten packing gland to proper setting.
- Replace bearing oil, or lubricate bearings with grease.
- Ensure pump shaft turns freely without noticeable dragging.
- Start pump and check for normal operation.

Internal combustion engines

- Remove all tape from sealed openings.
- Open fuel tank valve; shut water drain cocks and add coolant.
- Check oil drain plug; replace oil filter and add correct amount of oil to engine.
- Remove spark plugs and spray cylinder walls with a light engine oil.
- Replace spark plugs and crank engine several revolutions by hand to spread oil on cylinder walls.
- Lubricate all engine accessories.

- Where a distributor is used, clean inside and outside of cap; inspect cap and rotor for cracks; lubricate distributor sparingly with suitable lubricant; where a magneto is used, inspect breaker points for wear and gap; lubricate rotor.
- Where oil bath air cleaner is used, clean and fill with correct grade oil.
- Check all terminals and electrical connections.
- Start engine; run slowly for a few minutes; monitor oil pressure; if it fails to come up to correct reading, stop engine and investigate cause.
- Check oil level in crankcase and bring level up to proper mark on dipstick.

Electric motor

- Clean all debris accumulated during the storage period.
- Change motor bearing oil with special type of lubricant, do not overfill, use grease gun to lubricate bearings.
- Change oil in reduced voltage starters.
- Check that motor ventilation vents are open; clean dust and dirt from all moving parts of motor and panel.
- Check and tighten all electrical connections, replace overheated connections with new material; test all coils and heaters for continuity and shorts; clean all magnet surfaces; check for spare fuses of proper size; ensure all conduits or shielded cables are in good condition; check that all conduct points are corrosion free.
- Ensure service cabinet interior is moisture free.
- Operate all moving parts by hand before applying power.

CONCLUSION

Through investment in equipment for improved irrigation techniques, farmers expect to save considerable amounts of water, to increase yields and to improve crop quality.

Professionals and irrigation extensionists in association with manufacturers and farmers have been working for years on the proper O&M of irrigation system installations. Water conservation is and will continue to be a major goal for farmers, industry and governments. All parties concerned should cooperate to achieve this goal.

CHAPTER 21: Irrigation terminology

A.	
Actual evapotranspiration	Represents the actual rate of water uptake by the plant which is determined by the level of available water in the soil.
B.	
Balance of water resources and needs	The usable water resource of a certain water management unit in a given period of investigation, and the assessment and comparison of quantitative and qualitative characteristics of the water requirements to be supplied by this resource.
Basin irrigation	A gravity surface irrigation method in which crops are surrounded by a border to form a submersion check called basin of round, square or any other form. Irrigation water generally comes directly from the supply ditch/canal or from other basins.
Border irrigation	A sub-system of controlled flood (surface) irrigation in which the land is divided into parallel border strips demarcated from one another by earth ridges. Water is successively delivered into each strip from a head or field ditch at its upper end. On the upstream part of each strip is a flat zone, the level portion from which the stream of water spreads evenly across the entire downstream portion.
Bulk density	Bulk density or volume weight or apparent density or apparent specific gravity (As) of a soil is the dry weight of a unit volume of soil, which includes both the soil particles and the pores between them. It is expressed in g/cm^3 and varies from soil to soil according to texture and structure. It depends on soil porosity. Then, the larger the pore percentage the smaller the volume weight (Bulk density) of the soil.
C.	
Capacity of a well	The rate at which a well will yield water, in litres per second or cubic metres per hour.
Capital cost	The total expenditure incurred on a work since the beginning of its construction or supply of equipment and installation, excluding cost of operation, maintenance and repairs, but including cost of investigations and of all extensions and improvements.
Catchment area	The area from which a lake, a reservoir or a chosen cross-section of a stream or waterway receives water (= watershed or drainage basin, but usually smaller).
Command area (=designed area)	The specific land area, designed for irrigation by the irrigation system.
Centre pivot sprinkler	A sprinkler system in which the water source is in the centre, and a system of pipes and sprinkler heads rotates or pivots about the central point to water a given circular area.
Conventional technology	Technology based on a long history of experience without making use of later developments (compare with alternative technology).

Conveyance losses	Losses of water in transit from the source of supply to the point of service whether in natural channels or in artificial ones, such as canals, distributaries, ditches or watercourses. They comprise evaporation from the water surface, seepage, and incidental transpiration by vegetation growing in the water or along the banks of natural channels, canals or watercourses (= transmission losses).
Conveyance structures	Structures built to help provide general control and conveyance of the flow from the intake structures to the area to be irrigated.
Crop water requirement	The total water needed for evapotranspiration from planting to harvest for a given crop in a specific climatic regime when adequate soil water is maintained by rainfall and/or irrigation so that it does not limit plant growth and crop yield.
D.	
Discharge	Quantity (volume) of water passing through a given section of pipe, canal, valve, sprinkler nozzle or emitter etc. during a given period of time expressed in m^3 per hour, litres per second, gallons per minute, etc.
Drip irrigation	In its simplest form, it is an irrigation method using a system of perforated plastic pipes along the ground at the base of a row of plants (= trickle irrigation). In its more advanced form, it is a micro-irrigation system in which water flow is very low, generally less than 8 litres/h and without pressure, i.e. drop by drop. The water emerging infiltrates directly into the soil where it wets a volume of soil called bulb.
Drip tapes	Drip Irrigation Tapes are thin walled (0.12–1.1 mm) integral drip lines with emission points spaced at 10, 20, 30, 35, 45 cm, or any other distance apart, delivering very low quantities of water, 0.4–1.0 litre per hour at low operating pressures of 0.6–1.0 Bar. They are made of black PE (Polyethylene) in various diameters from 12 mm to 20 mm and in several wall thicknesses.
Drippers	Small emitters made of durable plastic, mounted on, or built in the small size (12–25 mm) PE irrigating pipelines at desired frequent spacing. The water enters at certain operating pressure and is discharged at zero pressure in the form of drops at constant low rates (1–24 litres per hour).
Duration of application	Time required for the completion of one irrigation cycle.
E.	
Economic value of unit of irrigation water	The value of a crop raised by a unit of irrigation water if run continuously throughout the life of the crop.
Effective root depth (D)	Soil depth from which the plants take nearly 80 percent of their needs in water (mostly from the upper part where root system is denser).
Erosion control	The application of necessary measures to control accelerated erosion of land surfaces by vegetation or artificial structures, such as terraces, dams or bunds.
Eutrophication	The process of a water body becoming anaerobic, i.e. without oxygen. Human activities that add nutrients to a water body can accelerate this process.
Exchangeable sodium percentage (ESP)	The degree of saturation of the soil exchange complex with sodium. It may be calculated by the formula: ESP = exchangeable sodium (meq/100 g soil)/cation exchange capacity (mec/100 g soil) x 100.

F.	
Fertigation	The fertilizers are applied through the system with the irrigation water, directly to the region, where most of the plants roots developed. The process, called "fertigation", is done with the aid of Fertilizer apparatus (injectors) placed at the Head Control unit of the system. The fertilizer, liquid or dry, is firstly dissolved and diluted in a separate container and then poured into the injector's tank, to be ejected into the system while in operation.
Field capacity (FC)	The amount of moisture retained in the soil one to four days after saturation, when the gravitational water is drained down to the lower soil layers. In light soils the time needed for the gravity water to drain is from 18 to 36 hours after saturation and in heavy soils from 36 hours to 4 days. In light soils the FC is lower than in heavy soils. In order to bring the state of moisture in the soil at FC at a certain depth, the amount of water required in a light soil is less than in a heavy soil. Again the same amount of water applied by irrigation or rain will wet a light soil to a greater depth than a heavy soil.
Flood irrigation	All types of irrigation which make use of rising water from flood for inundating areas without major structural works, e.g. flood recession, spate irrigation and wild flooding.
Flow-duration curve	A duration curve of stream flow, used for example to define minimum discharge and identify low-flow periods for the appraisal of irrigation water withdrawal.
Flow capacity	Water flow in m^3/hr or l/sec (lps) given or designed to fulfill the irrigation requirements of the command area at peak water demand. It is inversely proportional to the duration of application. It is often designed to be the lowest permissible to economize on sizes of pipes and equipment of the system network.
Friction losses (= head or pressure loss)	Loss of pressure (head) in the irrigation system that occurs during the flow of water in the system closed piping network as a result of the friction between the water and the pipes walls. The losses are proportional to the flow (discharge) and are dependent on the area of the pipe and on various obstructions to the flow such as contractions, outlets, valves, etc. The losses are measured in meters/feet or atmospheres/bars.
Fully automatic irrigation system	An irrigation system or network on a farm, whereby the water requirements of the plants are met automatically. It makes use of devices which measure soil moisture (e.g. tensiometer), or other indicators of irrigation need (e.g. time elapsed since rainfall), and trigger a series of operations to convey the necessary water through the network at the proper time.
Furrow irrigation	A method similar to corrugation irrigation used in permeable soils. It consists in feeding narrow furrows very close to one another with small discharges so as to wet more easily all the soil situated between two rows of crops (often orchards). Furrows parallel to the rows may be laid mechanically with a drill plough.
G.	
Gravity irrigation	Method of operating a system or part of a system using gravity alone, water being available at a sufficient level (or pressure) to ensure its conveyance or delivery to the fields or its distribution in the fields.
H.	
Hydraulic conductivity	1. The rate of flow of a fluid through a unit cross-section of a porous mass under a unit hydraulic gradient, at a specified temperature (sometimes called unit of permeability, transmission constant or coefficient of transmission). 2. The flux of water per unit gradient of hydraulic potential.

I.	
Individual irrigation system	Systems located downstream of the outlets served by the collective irrigation system and meant to deliver water to the farms or fields of an individual area.
Infiltration Rate	Infiltration or intake rate of soils is the maximum amount of water per unit of area, which can enter the soil (level) surface per unit time during the irrigation. It is expressed in millimeter per hour and it is governed by the conditions of the soil surface and the soil physical characteristics.
Irrigation Application Efficiency (Ea)	Percentage of the irrigation water, applied to the command area that is stored in the root-zone directly available to the crop; it is expressed in Percentage % or in fraction: Ea (%) = (water stored x 100)/ water applied
Irrigation cycle	Successive deliveries of water on all the units of a network in such a way as to achieve a given irrigation on the entire field concerned.
Irrigation efficiency	The ratio or percentage of the irrigation water consumed by crops of an irrigated farm, field or project to the water diverted from the source of supply. It is called farm irrigation efficiency or farm delivery efficiency when measured at the farm head-gate; field irrigation efficiency when measured at the field or plot; and water conveyance and delivery efficiency, or overall efficiency when measured at the source of supply.
Irrigation potential	Total possible area brought under irrigation, plus that which can be planned for irrigation in a river basin, region or country, from available water resources, with designs based on good technical practice at the time of assessing the potential.
Irrigation requirements	The quantity of water exclusive of precipitation, i.e. quantity of irrigation water, required for normal crop production. It includes soil evaporation and some unavoidable losses under the given conditions. It is usually expressed in water-depth units (millimetres) and may be stated in monthly, seasonal or annual terms, or for a crop period.
Irrigation water quality table	This indicates guidelines for the interpretation of water quality for crop production. The table was adapted from the University of California Committee of Consultants, the United States, in 1974 and revised in 1979. It emphasizes the long-term influence of water quality on crop production and farm management.
L.	
Leaching requirement	The fraction of water entering the soil that must pass through the root zone in order to prevent soil salinity from exceeding a specified value. Leaching requirement is used primarily under steady-state or long-term average conditions.
Localised Irrigation	The irrigation methods, where the water is delivered to the plants without spreading it over the entire area, but applied it to limited soil surface around the plants.
M.	
Main drainage system	System which conveys drainage water from the field drainage system to an outlet.
Method of water delivery	Way of making an irrigation system function to convey water from the source of supply to each field served by the system.

Micro-irrigation with mini-diffusers	A micro-irrigation system in which water is emitted in small sprinklings through fixed small diffusers in the form of fine droplets distributed over a certain area, or by individual low pressure jets localizing the water on the soil in separate spots. Their discharge is generally limited (20–60 litres/h at 1 bar) and often emitted in the form of circular sectors either to avoid wetting the neck of the trees or to limit the range on the sides of the space between rows, which should remain dry. Their use is limited to orchards.
Micro-emitters	Small water emitters (drippers, sprayers, bubblers, mini-sprinklers) made of durable plastic material. The rate of water discharge is in the range of 1 litre/hour to 170 litres/hour approximatively at operating pressure ranging from 0.6 bars to 2.0 bars. They have small water-passage diameters and the filtration requirements are from 100 microns (drippers) to 200 microns (minisprinklers).
Mobile micro-irrigation	An irrigation machine (generally frontal nozzle-line) in which the mobile nozzle-line functioning at low pressure applies water directly to the space between rows of annual crops. Suspended flexbles pipes fitted with mouthpiececs at their end feed continuously into small basins dug beforehand or simple partitioned corrugations.
Moisture (water) holding capacity	The quantity of water (moisture) hold by the soil is expressed by the "dry weight" percentage water in the soil $$PW = \frac{Ww - Wd}{Wd} \times 100$$ Where, **PW** is the soil water content percentage (dry weight), **Ww** is the weight of wet soil and **Wd** is the weight of the dry soil.
N.	
Net irrigable area	The total area within the extreme limits set for irrigation by a project, supply system or canal less areas excluded because of their un-suitability for irrigation (nature of the soil, ground too high to be irrigated by gravity flow or economically by pumps or other water lifting devices).
Net irrigation requirement	This is the crop's irrigation need (without including losses of any kind) expressed as a layer of water in millimeters per day, month or other period of time.
Nominal discharge of a dripper	Discharge in litres per hour at the nominal pressure indicated by the manufacturer. This discharge is determined by a test carried out as per the ISO standard on 25 samples taken at random. In the case of a self-regulating dripper, the test pressure is the arithmetic mean of the minimum and maximum pressures in the regulation range indicated by the manufacturer.
O.	
Operation and maintenance (O&M)	Operation is the organized procedure for causing a piece of equipment, a treatment plant, or other facility or system to perform its intended function, but not including the initial building or installation of the unit. Maintenance is the organized procedure for keeping the equipment, plant, facility or system in such condition that it is able to perform its intended function continually and reliably.
Overall efficiency	The ratio or percentage of the irrigation water consumed by crops to the water diverted from the source of supply (measured at the source of supply).
Overhead irrigation	Irrigation by which water is ejected into the air to fall on the soil surface as spray.
Over-irrigation	Excessive irrigation with regard to the actual requirements, due to excessive doses of watering, an insufficient irrigation interval or an overestimation of the requirements (lesser evapotranspiration or excess of rains with respect to the normal). It causes either a leaching of the soil if it is sufficiently drained, or a water-logging of the soil which harms crop growth.

P.	
Pan evaporation	Rate of water loss by evaporation from an open water surface of pan (usually, Class A pan or Colorado sunken pan).
Peak period crop water requirements	For a given crop, the peak crop water requirements during the month of highest water requirements.
Percolation rate	The maximum rate at which water will flow into the subsoil from the topsoil under specific conditions, expressed in millimetres per hour or day.
Perennial irrigation	An irrigation is termed perennial when the lands of the area can be irrigated throughout the year and have the volume of water actually required.
Perforated pipe sprinkler irrigation	A sprinkler method in which the nozzle-lines consist of portable and lightweight pipes, the wall of which is perforated with several rows of small holes in such a way as to cause the water to be applied on both sides of the nozzle-line.
Permanent wilting point	The moisture content of the soil, expressed as a percentage of the soil volume or as a percentage of dry weight, at the time when the leaves of a plant growing in the soil first undergo a permanent reduction in their moisture content as the result of the deficiency in the soil moisture supply.
Permissible velocity	The highest velocity at which water may be carried safely in a canal or other conduit. The highest velocity throughout a substantial length of a canal or other conduit that will not scour.
Poor drainage	Occurs in soils which lose gravitational water slowly, or which are situated where the groundwater table remains high in the profile. In most years, the soil root zone loses excess soil water only during the summer months. In an unimproved condition, successful cropping is unlikely (e.g. standing water, water margin, wetland and peatland environments).
Pore Space of Soil	Percentage of volume of the soil not occupied by the soil particles and filled with water and air. Fine texture soils (heavy) have greater pore space than coarse texture (light) soils. The pores can be divided theoretically into capillary pores holding the water against gravitational pull and non-capillary pores containing air. Downward movements of water due to gravity are through these pores.
Potential evapotranspiration (ETo)	1. The amount of water that could pass into the atmosphere by evapotranspiration if the amount of soil water were not a limiting factor. 2. The amount of water utilized by a crop for its growth plus evaporation from the soil if the soil contains sufficient moisture for crop growth at all times.
Potential yield (of a well)	The greatest rate of artificial withdrawal from an aquifer which can be maintained throughout the foreseeable future without regard to cost of recovery. The potential yield (or physical yield limit) is, therefore, equal to the present recharge, or that anticipated in the foreseeable future, less the unrecoverable natural recharge.
Pressure of the system	Maximum water pressure or head of water needed for the normal operation of the system and encompasses: a) the total losses of head due to friction in the pipes, the connector fittings and other accessories from the beginning to the distal end of the close piping network, b) the pressure required for the water emitter, c) the pressure needed for the head control unit, d) plus or minus the difference in elevation from the beginning to the distal end of the close piping network pressure.
Pumping irrigation	Method of operating a system or part of a system using, fully or partly, an artificial pressure for ensuring the conveyance of water, its delivery or distribution in the fields.

R.	
Rainfall intensity	The rate at which rainfall occurs expressed in depth units per unit of time. It is the ratio of the total amount of rain to the length of the period in which the rain falls.
Readily available moisture	The state of moisture in the soil, which amounts 40 to 70 percent of the total available moisture (Sa) easily absorbed by the plants. It is the product of the Sa multiplied by P (fraction) maximum permissible moisture deficit or depletion of the Sa in percentage, hence: Readily available moisture = Sa *x*P
Reference crop evapotranspiration (ETo)	The rate of evapotranspiration from an extensive surface of 8–15 cm tall green grass cover, actively growing, completely shading the ground and not short of water. Alternative approaches for estimating ETo are the radiation, the Penman and the Pan Evaporation (presented in FAO bulletin Irrigation and Drainage No 24). In all methods is expressed in millimeter per day mean value over 30 or 10 days period. ETo data are normally available in all countries. They can also be computed from climatic data.
Regulation structure	A stage-discharge regulating device of a spillway. It may be of any form, viz. weir, side channel, glory spillway, orifice, tube, pipe or a channel. (= control structure)
Regulation with downstream control	Method of regulation in which the flow in a canal (or in a pipeline) is controlled at a gate by the level of the water (or pressure) measured by a sensor or by a float connected to the gate placed in the immediate **downstream** of the gate. It is a **delivery**-oriented control method.
Regulation with upstream control	Method of regulation in which the flow in a canal (or in a pipeline) is controlled at a gate by the level of the water (or pressure) measured by a sensor or by a float connected to the gate placed in the immediate **upstream** of the gate. It is a **supply**-oriented control method.
Roll-move sprinkler lateral system	A sprinkler method in which the nozzle-line, which carries medium pressure sprinklers, is used as an axle to the wheels which support it at regular intervals. Watering is done in a permanent shift and the nozzle-line is moved manually between waterings to its new position by rolling it fully.
S.	
Salinity control	Abatement or prevention of saltwater contamination of agricultural, industrial and municipal water supplies, or reducing alkaline salts and preventing deterioration of cultivable lands.
Seasonal irrigation	Irrigation is termed seasonal when the lands of the area are irrigated only during a part of the year, called watering season.
Semi-automatic field water distribution system (partially automatic system)	Irrigation system in which the water distribution and field application are partly automatic and partly manual. A semi-automatic system may carry out a sequence of operations automatically for a single irrigation, but need to be manually started or manually reset prior to the subsequent irrigation. It may involve use of volumetric or timer controlled valves that are started manually but which close automatically.
Semi-module (flexible module)	A device that automatically delivers a discharge which is independent of fluctuations of water level or pressure on the delivery side, and only varies with water level or pressure on the supply side (used for regulation with down-stream control).
Sensitivity analysis	The study of the influence of discrete parameter changes on optimized results. Those parameters whose changes in value have more significant influence on the results need treating with great care, while other parameters can be recognized as relatively insignificant.

Social benefits	Benefits as a result of the project, during and after construction, consisting mainly of opportunities for: (i) employment of labour; and (ii) employment of capital.
Sodium adsorption rate (SAR)	A ratio for soil extracts and irrigation waters used to express the relative activity of sodium ions in exchange reactions with soil: $SAR = Na^{+} x [(Ca^{2+} + Mg^{2+})/2]^{-0.5}$ where the ionic concentrations are expressed in meq/litre.
Soil available moisture (Sa)	Available moisture (or water) is the percentage content of water in the soil at the range between Field Capacity and Wilting Point. It is the principal source of water for plants and is usually called as capillary water. Available Water Sa % = FC% – WP% The "dry weight" percentage water in the soil at any state is converted to "volume weight" when multiplied by the soil Bulk density. Values of Available water are always given in "volume weight". Then, Sa % "dry weight" x Bulk density g/cm^3 = Sa % "volume weight" The percentage "volume weight" Soil Available Moisture Sa can be expressed as amount (depth) of water in mm per meter soil depth, e.g. Sa 10.92 % volume weight = 109.2 mm of water per meter depth of soil.
Soil moisture characteristics	Saturation, field capacity and wilting point of a soil can be determined by several methods direct or indirect required field and or laboratory work and simple calculations. The "gravimetric method" to determine FC is the oldest one, the most common and reliable method. Soil samples taken from several depths are put in an oven at 105° Celsius and kept for 8 hours (light soils) to 16 hours (heavy soils) Three soil samples are weighted before and after drying and the difference in dry weight gives FC percentage by weight. Other methods to determine the Soil FC are a) by the use of pressure on the soil samples, 0.10–0.30 atmospheres in the "pressure plate apparatus", b) the "moisture equivalent centrifuge" and c) the "speedy moisture test". Figures given by the "pressure plate apparatus" method are higher than those given by the "gravimetric" method in heavy soils. The permanent wilting point can be determined by the "pressure membrane apparatus" method, where 15 atmospheres air pressure is applied to saturated soil samples and then by the gravimetric method the moisture content is found.
Soil moisture deficit	The amount of water that must be applied to the soil to cause thorough drainage.
Soil permeability	It is the readiness with which the soil conducts or transmits water. It is quantitatively defined as Hydraulic Conductivity (K) and greatly depends on the soil texture and the quality of the irrigation water.
Soil profile	The whole arrangement of different layers (horizons) from top to bottom, from soil surface to mother rock. The top layer, called "A" horizon, is normally cultivated for crops. The layer below is called horizon "B" usually with higher clay content and further below is horizon "C". Horizons "A" and "B" can be defined as the soil. Soils formed under different conditions have different profiles. Young soils formed by alluvial materials have moderate or no profile development and sometimes show a significant variation in texture within the depth of the root zone. An average depth of 70 cm and even less is suitable nearly for all kind of crops. Under modern irrigation and nutrient supply methods (fertigation) soil depths of 45 cm are sufficient for most vegetables and shallow root trees.

Soil saturation	When all the pore space of the soil is filled with water, after a heavy irrigation or rainfall, the soil is saturated. The amount of water, which can be held by a soil at saturation capacity, depends on the volume of its pore space. Hence the saturation capacity is larger in heavy soils than in light ones.
Soil texture	The soil as a physical body is described by the size and arrangement of its particles, determining the porosity. Soil particles are divided into three major size fractions: a) sand, b) silt and c) clay. Soils are classified, according to United States Department of Agriculture (USDA) classification system, by their textural class i.e. the percent composition of sand, silt and clay. Soils with high sand content are called "light soils" or "sandy soils" and soils with high clay content are called "heavy soils". The soil properties such as water holding capacity and intake rate mostly depend on the texture.
Soil water stress	The sum of soil water tension and osmotic pressure to which water must be subjected to be in equilibrium with soil water.
Soil moisture tension	The equivalent negative pressure or suction in the soil moisture; expressed in pressure units (bar or pascal).
Spate irrigation	A method of random irrigation using the floodwaters of a normally dry (stream, river) system. It includes the construction of earthen diversion banks across the bed and then canals leading to embanked fields where the water is ponded until total infiltration.
Sprinkler irrigation system	It is a designed network of pipes with sprinkler emitters or nozzles attached for shooting water jets or spraying water in the form of drops over the land surface, under pressure.
Subsurface drainage system	Any drainage system (drainage wells, open ditches or drain pipes) that is designed to control the groundwater table.
Supplemental irrigation	Irrigation carried out only occasionally to make good for short and irregular drought periods.
Surface drainage system	Shallow ditches or open drains that serve to receive surface flow or drainage water.
Surface irrigation	A method of irrigation in which water is applied to the land by allowing it to flow by simple gravity, before infiltrating. It includes various systems depending upon the relative magnitude of the surface flooding phase and infiltration phase after accumulation (submersion).
T.	
Technology transfer (transfer of know-how)	Technology transfer consists in supplying project users or training personnel with technical knowledge and training essential for proper command of O&M functions. This transfer may remain too theoretical or abstract if not accompanied by: a transfer of know-how from the development corporation officials to the users; a set of demonstrations; and a suitable follow-up of the concrete operations (technical and management).
Tensiometer	An instrument for measuring the suction that plant roots have to exert in order to extract moisture from the soil.

U.	
Undertree sprinkler method	Sprinkler method used in orchards with small sprayers with an outstretched jet in order not to wet the leaves and avoid the wind effect on the distribution of water. Such sprayers can be permanent, semi-permanent or portable.
W.	
Water conveyance and delivery efficiency	The ratio or percentage of the irrigation water delivered at the irrigation plot to the water diverted from and measured at the source of supply.
Water emitter	Also named "water distributor" is a device of any kind, type and size, which fitted on a pipe, is operated under pressure to discharge water in any form - by shooting water jets into the air (sprinklers and rain guns), by small spray or mist (sprayers), by continuous drops (drippers), by small stream or fountain (bubblers). The pressure/discharge rate relationship and other performances, e.g. coverage pattern, way of dissipating energy (pressure), size and type of connection, is always specified. Sprinklers, drippers, spitters, sprayers, bubblers, pulsators and garden hoses are water emitters.
Wetting front	The depth in the soil above that the moisture content of the soil is at maximal field capacity, is called the wetting front and depends on the amount of water applied and the soil texture.
Wilting point (WP)	The state of moisture in the soil at which water content is very low and not readily available to the plants. At "permanent" wilting point the moisture in the soil is even less, only hygroscopic water, and the plants wilt permanently. The water percentage content at wilting point is nearly half the water content at field capacity.

CHAPTER 22: Irrigation equipment supply database – IES

Irrigation Equipment Supply (IES) is a joint initiative of the Water Resources, Development and Management Service of FAO and the International Programme for Technology and Research in Irrigation and Drainage (IPTRID). It has been developed as part of FAO's mandate to provide information on irrigation and based on the first version of this handbook. Potential beneficiaries of IES are those who need to locate information on irrigation equipment at regional or country level.

IES database seeks to establish an up-to-date list of of irrigation equipment services and Suppliers/Manufacturers providing irrigation equipment worldwide. National Suppliers/Manufacturers can be displayed through the selection of the country on a map. Moreover, the website offers a database query facility for identifying Suppliers/Manufactures providing specific irrigation equipment through a small set of search criteria. Apart from obtaining detailed Supplier/Manufacturer contact information, a list of the irrigation equipment provided by each Supplier/Manufacturer is given. The equipement option provides a detailed description of equipment items, including a sample photo and a list of corresponding international standards, where available. Retrieval is facilitated by a name and/or keyword search option. The complete list of standard referring to irrigation equipment, organized by institution is available at the standards page.

This application is primarily Supplier/Manufacturer-driven in the sense that information contained in the database is inserted and maintained exclusively by the Suppliers/Manufacturers of irrigation equipment. Once the registration has been submitted, the Supplier/Manufacturer receives an e-mail confirmation with a confidential Supplier/Manufacturer identification code permitting an independent update of the Supplier/Manufacturer and equipment information at any time. FAO can thus not give any guarantee that the information is correct and up-to-date and does not accept any liability arising out of the information. Information submitted is visible in the database only after verification by the IES staff. FAO reserves the right to reject any non-pertinent contribution.

IES is located on the Web site (Figure 22.1):

http://www.fao.org/nr/water/ies/

FIGURE 22.1 - IES Web site.

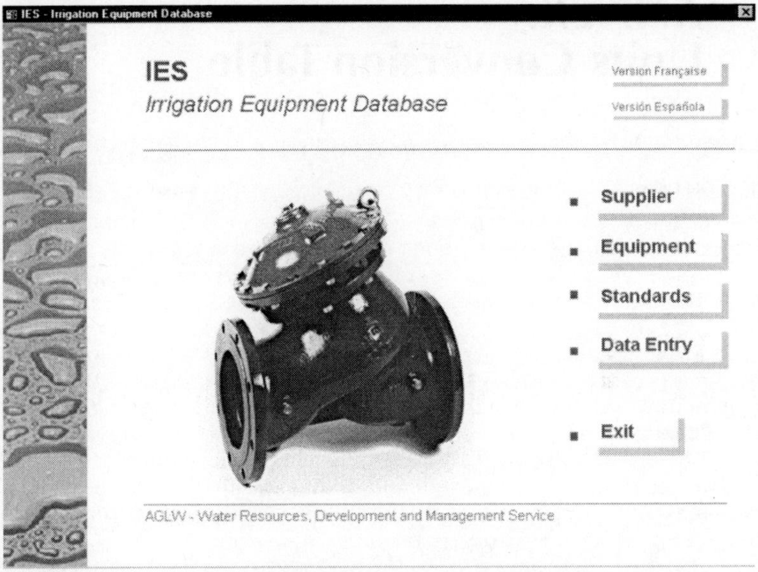

The mention of specific companies or of their products or brand names does not imply any endorsement by, and the view expressed do not necessarily reflect the views of, the Food and Agriculture Organization of the United Nations.

For any kind of technical query regarding this website, the functioning of the database application as well as any query regarding the content of IES, we propose you send a message to the administrators (ies@fao.org).

ANNEX: Units Conversion Table

UNITS CONVERSION TABLE

Old Units	SI (metric) Units	Convert	
A	B	A to B	B to A
Lenght		**Multiply by**	
inch	mm	25,4	0,0394
inch	m	0,0254	39,3700
inch	mil	1000	0.001
mil	mm	0.0254	39.37
ft (foot)	inch	12	0,0833
ft	cm	30,48	0,0328
ft	m	0,3048	3,2808
yard	m	0,9144	1,0940
Chain	m	20,1168	0,0497
mile	m	1,609	-
mile	km	1,6090	0,62137
mile (nautical)	km	1,852	0,5399
Area		**Multiply by**	
inch² (in²)	cm²	6,451	0,155
inch²	m²	0,000645	1 550,150
ft²	m²	0,0929	10,764
sq. mile	km²	2,5899	0,386
acre	m²	4 046,87	-
acre	ha	0,405	2,471
hectare (ha)	m²	10 000	0,0001
km²	ha	100	0,01
Mass Weight		**Multiply by**	
oz (ounce)	g	28,35	0,0353
lb (pound)	kg	0,4536	2,205
oke	kg	1,27	0,78737
cwt (Hundredweight)	kg	50,8	0,0197
long ton (U.K.)	kg	1 016,0	-
short ton (USA)	kg	907,0	-
metric ton	kg	1 000	-
Volume Capacity		**Multiply by**	
inch³ (in³)	cm³ (cc)	16,387	0,061
ft³	L (liters)	28,3	0,0353
ft³	m³	0,0283	35,315
m³	l	1 000	0,001
gallon (Imp.)	l	4,546	0,2199
gallon (Imp.)	m³	0,0045	220
gallon (USA)	l	3,785	0,264
gallon (USA)	m³	0,00378	264

1 Acre foot = 1234 m³

23.2 *Annex – Units Conversion Table*

UNITS CONVERSION TABLE			
Old Units	**SI (metric) Units**	**Convert**	
A	B	A to B	B to A
Flow		**Multiply by**	
Imp. gal./min. (IGPM)	l/s	0,0758	13,199
Imp. gal./min. (IGPM)	l/h	272,88	0,00366
Imp. gal./min. (IGPM)	m^3/h	0,273	3,666
U.S. gal./min.	l/s	0,0631	15,85
U.S. gal./min.	l/h	227,12	0,0044
U.S. gal./min.	m^3/h	0,227	4,4
ft^3/s (CFS)	l/s	28,32	0,0353
ft^3/s (CFS)	m^3/h	101,94	0,0098
Pressure		**Multiply by**	
lb/in^2 (psi)	m water column	0,703	1,465
lb/in^2 (psi)	Bar	0,06895	14,5
lb/in^2 (psi)	kPa	6,89	0,145
kg/cm^2	m water column	10	0,1
kg/cm^2	Bar	0,981	1,0193
kg/cm^2	Atm	0,9678	1,0332
Atm.	kPa	101,3	0,00987
Atm.	kN/m^2	101,3	0,00987
Atm.	kN/m^2	0,1	10
Atm.	Bar	1,013	0,9869
Atm.	m water column	10,33	0,0968
Atm.	mm mercury column	760	0,001316
Bar	kPa	100,0	0,01
Bar	m water column	10,19	0,098
Velocity		**Multiply by**	
mph	km/h	1,609	0,6215
mph	m/sec	0,4469	2,2374
Heat		**Multiply by**	
Btu	kcal	0,252	3,968
Btu	kJ	1,055	0,9478
kcal	kJ	4,18674	0,2388
Horsepower (HP)		**Multiply by**	
hp	kW	0,7457	1,341
Electrical Conductivity		**Multiply by**	
mmhos/cm	dS/m	0,99999	0,99999